AF606217

90 0593918 5

DIFFUSION AND TRANSPORT OF POLLUTANTS IN ATMOSPHERIC MESOSCALE FLOW FIELDS

ERCOFTAC SERIES

VOLUME 1

Series Editors

Aims and Scope of the Series

ERCOFTAC (European Research Community on Flow, Turbulence and Combustion) was founded as an international association with scientific objectives in 1988. ERCOFTAC strongly promotes joint efforts of European research institutes and industries that are active in the field of flow, turbulence and combustion, in order to enhance the exchange of technical and scientific information on fundamental and applied research and design. Each year, ERCOFTAC organizes several meetings in the form of workshops, conferences and summerschools, where ERCOFTAC members and other researchers meet and exchange information.

The ERCOFTAC Series will publish the proceedings of ERCOFTAC meetings, which cover all aspects of fluid mechanics. The series will comprise proceedings of conferences and workshops, and of textbooks presenting the material taught at summerschools.

The series covers the entire domain of fluid mechanics, which includes physical modelling, computational fluid dynamics including grid generation and turbulence modelling, measuring-techniques, flow visualization as applied to industrial flows, aerodynamics, combustion, geophysical and environmental flows, hydraulics, multi-phase flows, non-Newtonian flows, astrophysical flows, laminar, turbulent and transitional flows.

Diffusion and Transport of Pollutants in Atmospheric Mesoscale Flow Fields

Edited by

Albert Gyr

and

Franz-S. Rys

Institut für Hydromechanik und Wasserwirtschaft,
Eidgenössische Technische Hochschule Zürich,
Switzerland

KLUWER ACADEMIC PUBLISHERS

DORDRECHT / BOSTON / LONDON

Library of Congress Cataloging-in-Publication Data

Diffusion and transport of pollutants in atmospheric mesoscale low fields / edited by Albert Gyr and Franz-S. Rys.
p. cm. -- (ERCOFTAC ; 1)
Includes bibliographical references and index.
ISBN 0-7923-3260-1 (acid-free)
1. Air--Pollution--Congresses. 2. Mesometeorology--Congresses. I. Gyr, Albert. II. Rys, Franz-S. III. Series: ERCOFTAC (Series) ; 1.
TD883.1.D54 1995
628.5'3--dc20 94-39300

ISBN 0-7923-3260-1

Published by Kluwer Academic Publishers,
P.O. Box 17, 3300 AA Dordrecht, The Netherlands.

Kluwer Academic Publishers incorporates
the publishing programmes of
D. Reidel, Martinus Nijhoff, Dr W. Junk and MTP Press.

Sold and distributed in the U.S.A. and Canada
by Kluwer Academic Publishers,
101 Philip Drive, Norwell, MA 02061, U.S.A.

In all other countries, sold and distributed
by Kluwer Academic Publishers Group,
P.O. Box 322, 3300 AH Dordrecht, The Netherlands.

Printed on acid-free paper

Printed in the Netherlands

Contents

Preface

In regions as densely populated as Western Europe predictions of ecological implications due to the transport of pollutants in the atmosphere of medium range distance are important for two main reasons: (1) Evaluation of a possible influence of existing or planned sources of pollutants on the environment. (2) Minimization of damages in case of accidents. It is evident that for (1) a large amount of numerical simulations is necessary whereas for (2) the time available for a prediction and warning of imminent danger is extremely short and of utmost importance. It can be concluded that in both cases a solution is dependent on high-speed computations. Computations of this kind are costly. Therefore one of the goals is to lower the costs by meaningful cooperation between the different groups in Europe working on such problems and using unified data banks. As a first step in this direction the "European Research Community On Flow, Turbulence And Combustion" (ERCOFTAC) has sponsored a series of courses and meetings dealing with the transport of pollutants in the atmosphere.

From 23 to 27 August 1993 an International Summerschool on Diffusion and Transport of Pollutants in Atmospheric Mesoscale Flow Fields was held in Manno near Lugano. It was followed by a Workshop (WS) on Intercomparison of Advanced Practical Short-Range Atmospheric Dispersion Models.

In principle, this monograph contains the material presented in a series of invited talks given at this Summerschool (SS) in the CSCS (Centro Svizzero di Calcolo Scientifico) in Manno. This SS was the second in a series of three meetings in the frame of the "European Harmonization in Atmospheric Dispersion Systems for Regulatory Purposes". The first WS on "Objectives for next generation of practical short-range atmospheric dispersion models" was held at RISO Nat. Lab. Roskilde, Denmark, May 6- 8, 1992, and the third WS on "Application of practical atmospheric dispersion models to industrial sources-intercomparison, administration aspects and implications at the EC level" is scheduled to take place in Mol, Belgium in 1994.

Following the desire of all participants to disperse of a standard textbook, the whole set of contributions was compacted and regrouped into 8 chapters. Where necessary the mathematical part is explicit. It is aspired to present, in this formal form, a pedagogic frame of a didactic introduction for the expert as well as for any newcomer interested in one of these fields.

In chapter 1 an introduction to the basics of fluid dynamics of the atmosphere and the local events and meso-scale processes is given. These topics were presented at this SS by B.W. Atkinson.

Next, the type of PDE's of limited area models for atmospheric flows, the problem of appropriate boundary conditions describing the topographical constraints and the well posedness are discussed (D. Eppel and U. Callies in Ch. 2).

The thermodynamics of the atmosphere, in general the dry and wet atmosphere, its stability and radiation processes, budgets and the influence of their sum is discussed by M. Beniston and J. Schmetz in Ch. 3.

Scaling and similarity laws for stable and convective turbulent atmospheric boundary layers and the influence of inhomogeneous terrain on the advection and the vertical dispersion are, together with the method of large eddy simulation (LES), which allows to calculate these atmospheric flows, discussed by F.T.M. Nieuwstadt in Ch. 4.

An introduction to statistical approaches in turbulent dispersion, turbulent diffusion of particle and chemical reactions in fluxes are treated by H. van Dop in Ch. 5, whereas a review of various theoretical modelings of diffusion and dispersion of pollutants and pollutant gases follows in Ch. 6 (by T. Mikkelsen). In Ch. 7 P.G. Mestayer and S. Anquetin discuss the influence of urban heat production (and pollution) on the local developments in climate.

Finally, E. Fedorovic discusses two examples of atmospheric inversion layers with observed negative temperature gradient and attempts to describe the lapping inversion, the stable boundary layer and the nocturnal inversion in terms of simple parametricable models.

We hope that this volume will enhance the basic scientific research activity, on the one hand, and on the other, stimulate ecologically motivated applications on the broadest possible basis, thus obtaining a better control and an efficient reduction of the harmful effects of air and water pollution and aiding in eliminating hazardous health effects on mankind and avoiding the destruction of the sensible ecosystem of Mother Nature.

Albert Gyr
Franz-S. Rys

Zurich, March 1994

Acknowledgements

First of all, we would like to thank the three institutions which made this Summerschool possible; the Joint Research Centre in Ispra and especially Dr. C. Cuvelier for the coordination of the contributions to the simulation of different atmospheric transport problems made by the different European groups. We would like to thank the CSCS in Manno, who, as our host, put their lecture rooms and computational facilitiesat our disposal. We are especially grateful to Dr. R. Gruber for his assistance during the course, and to the ETHZ as the main organizing institution and to Prof. Dr. Th. Dracos in particular for his active support which made this Summerschool possible.

Naturally, a Summerschool lives from the quality of its teachers. We would like to thank all the active contributors for their excellent work and their cooperation in producing this book.

Last but not least, our thanks goes to Dr. Studerus, who did the enormous work of re-formating the various texts written in so-called "compatible" formats into a uniform camera-ready product.

One of us (A.G.), in addition, would like to thank his daughter, Nadia, for her wonderful help as the secretary of the Summerschool.

List of Authors & Addresses

B.W. Atkinson, Queen Mary & Westfield College, University of London, England

D. Eppel & U. Callies, GKSS Forschungszentrum, Geesthacht/ b. Hamburg, Germany.

M. Beniston, Dept. of Geography, ETH-Zürich, Switzerland

J. Schmetz, European Space Operation Center/ESOC, Darmstadt, Germany

F.T.M. Nieuwstadt, Lab. Aero Hydrodynamics, J.M. Burgers Centre, Delft, The Netherlands

H. van Dop, Utrecht University, Utrecht, The Netherlands

T. Mikkelsen, Dept. of Meteorology & Wind Energy, Riso Nat. Lab., Roskilde, Denmark

P.G. Mestayer & S. Anquetin, Lab. de Mécanique des Fluides, Ecole Centrale de Nantes, France

E. Fedorovic, Lab. de Mécanique des Fluides, Ecole Centrale de Nantes, France

I Introduction to the fluid mechanics of meso-scale flow fields

B W Atkinson

1.1 Introduction

Meso-scale airflows are essentially characterised by a horizontal length scale of a few tens of kilometres. Many of them, particularly those of importance to the distribution of pollutants, are relatively shallow (a few kilometres at most) and are significantly influenced by the nature of the underlying earth's surface. This chapter is concerned with clarifying the term 'meso-scale' and presenting the basic fluid mechanics of meso-scale airflows, bearing in mind the above characteristics. It starts with a consideration of the observational and theoretical background to the idea of a meso-scale of atmospheric flows and then investigates the equations that aid understanding of the mechanics of these flows. Presentation of the basic equations is followed by elucidation of how they may be averaged, thus revealing the turbulent transfer terms. This is followed by a consideration of four approximations that may be made in the momentum and continuity equations, identifying those of most importance to meso-scale airflows. The need for and procedures involved in specifying turbulence closure follows. Much of the analysis of meso-scale airflows now involves solution of the momentum and other conservation equations in numerical models, so the notions of a model domain, grid volumes and sub-grid scale processes underlie the treatment in this chapter. Readers will find of value the books by Garratt (1992) and Stull (1988) for full treatments of the atmospheric boundary layer and that by Pielke (1984) for its treatment of the modelling of meso-scale airflows.

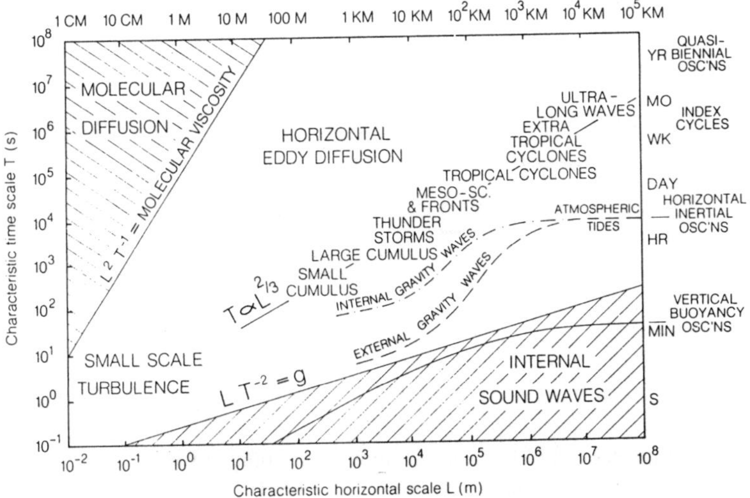

Figure 1. Spatial and temporal characteristics of atnospheric phenomena (after Smagorinsky 1981).

A. Gyr and F-S. Rys (eds.), Diffusion and Transport of Pollutants in Atmospheric Mesoscale Flow Fields, 1–22.

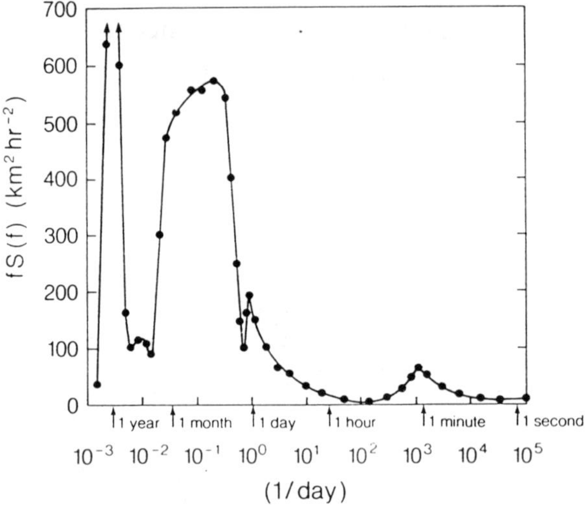

Figure 2. Average kinetic energy of the zonal wind component in the free atmosphere (after Vinnichenko 1970).

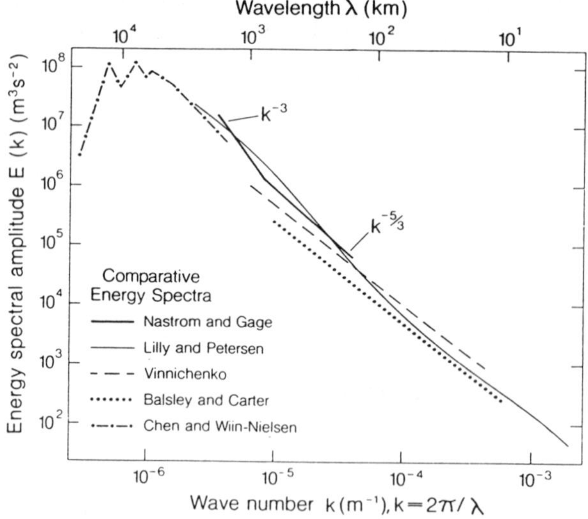

Figure 3. Five spectra of horizontal winds over a range of wavelength from about five kilometres to the earth's circumference (after Lilly 1983).

1.2 The meso-scale

The term 'meso-scale' was probably first used in the English-language meteorological literature by Ligda (1951) in the context of a review of radar meteorology. Radar was able, for the first time, to monitor atmospheric phenomena of sizes greater than those visible from the conventional surface weather station and smaller than those resolved by the analysis of routinely gathered weather information (the synoptic or coriolis scale) – hence the descriptor middle or meso-scale This simple, yet powerful classification was further established by Fujita, Newstein and Tepper (1956) and Tepper (1959) who recognised its importance in weather forecasting.

1.2.1 Observation

It has subsequently become commonplace to recognise that the atmosphere comprises a suite of circulations which have a wide range of spatial and temporal dimensions . In spatial terms the horizontal dimension is the main discriminant and Figure 1 shows that atmospheric features range in size from a few millimetres to thousands of kilometres. Their associated temporal dimensions range from a few seconds to several months, possibly a few years. Early spectra of wind speeds appeared to confirm the clear three-fold division of the atmospheric motion field (Figure 2) but it is not so evident in more recent ones (Figure 3) Synthesis of the qualitative and quantitative observational evidence shows some common ground as to the meaning of meso-scale (Figure 4) with a range of 10 to 100 km

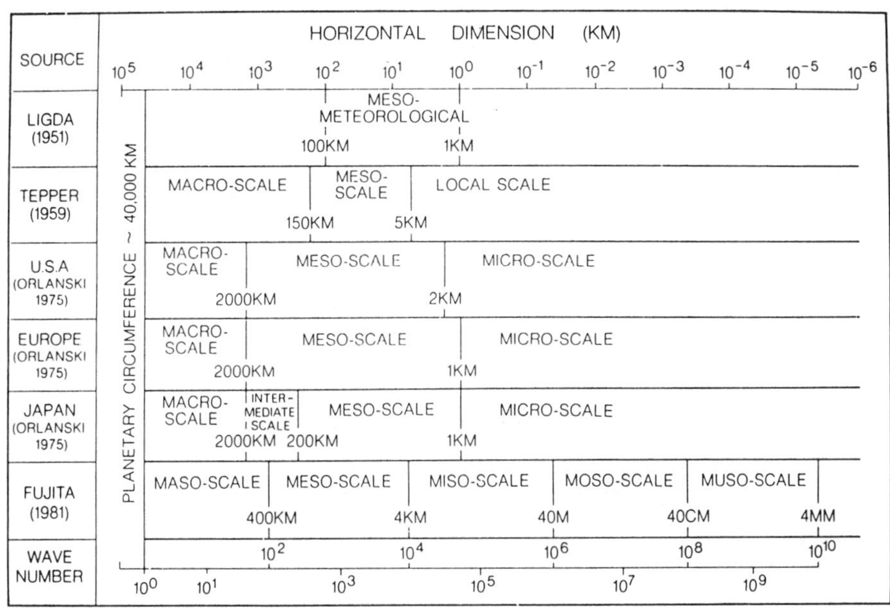

Figure 4. Non-dynamical scalings of atmospheric motion.

1.1.2.2 Theory

It is possible to acquire a more rational appreciation of scale by considering fundamental frequencies in the atmosphere. Three are readily identified: Brunt-Vaisala, Inertial and Planetary (Figure 5).

FUNDAMENTAL FREQUENCIES

1. BRUNT-VAISALA (N)	g ACCELERATION DUE TO GRAVITY
$N = \left(\frac{g}{\theta}\frac{\partial\theta}{\partial z}\right)^{1/2} \sim 10^{-2}s^{-1}$	θ POTENTIAL TEMPERATURE
	z VERTIVERTICAL DISTANCE
	Ω ANGULAR VELOCITY OF EARTH
2. INERTIAL (f)	ϕ LATITUDE
$f = 2\Omega\sin\phi \sim 10^{-4}s^{-1}$	U HORIZONTAL VELOCITY OF AIR
3. PLANETARY (P)	β RATE AT WHICH CORIOLIS PARAMETER (f) CHANGES WITH LATITUDE
$P = (U\beta)^{1/2} \sim 10^{-6}s^{-1}$	

Figure 5. Fundamental frequencies in the atmosphere.

These frequencies are associated with periods of minutes, hours and days respectively. The Brunt-Vaisala frequency is that expected for vertical oscillations in the stratified atmosphere. It is clearly related to 'cloud-scale' motions, such as can be seen every day in the sky. The inertial frequency results from the rotation of the earth. The planetary frequency results from the latitudinal variation of the coriolis parameter, the so-called beta-effect and hence influences the largest motions on the planet.

SCALE BY FREQUENCY (F)	
1. SMALL	$F > N$
2. MESO	$f < F < N$
3. SYNOPTIC	$P < F < f$
4. PLANETARY	$F < P$

Figure 6. Four scales of atmospheric motion defined by three fundamental frequencies. N is the Brunt-Vaisala frequency; f is the inertial frequency; P is the planetary frequency - as shown in Figure 5.

Figure 6 shows a four-fold hierarchy of scales based on the three fundamental frequencies, the meso-scale being bounded by the Brunt-Vaisala and inertial frequencies. This and other dynamically-based scalings (most notably the widely used one suggested by Orlanski (1975) (Figure 7)) appear in Figure 8, which shows a compromise classification in which the meso-scale ranges from 1 to 100km.

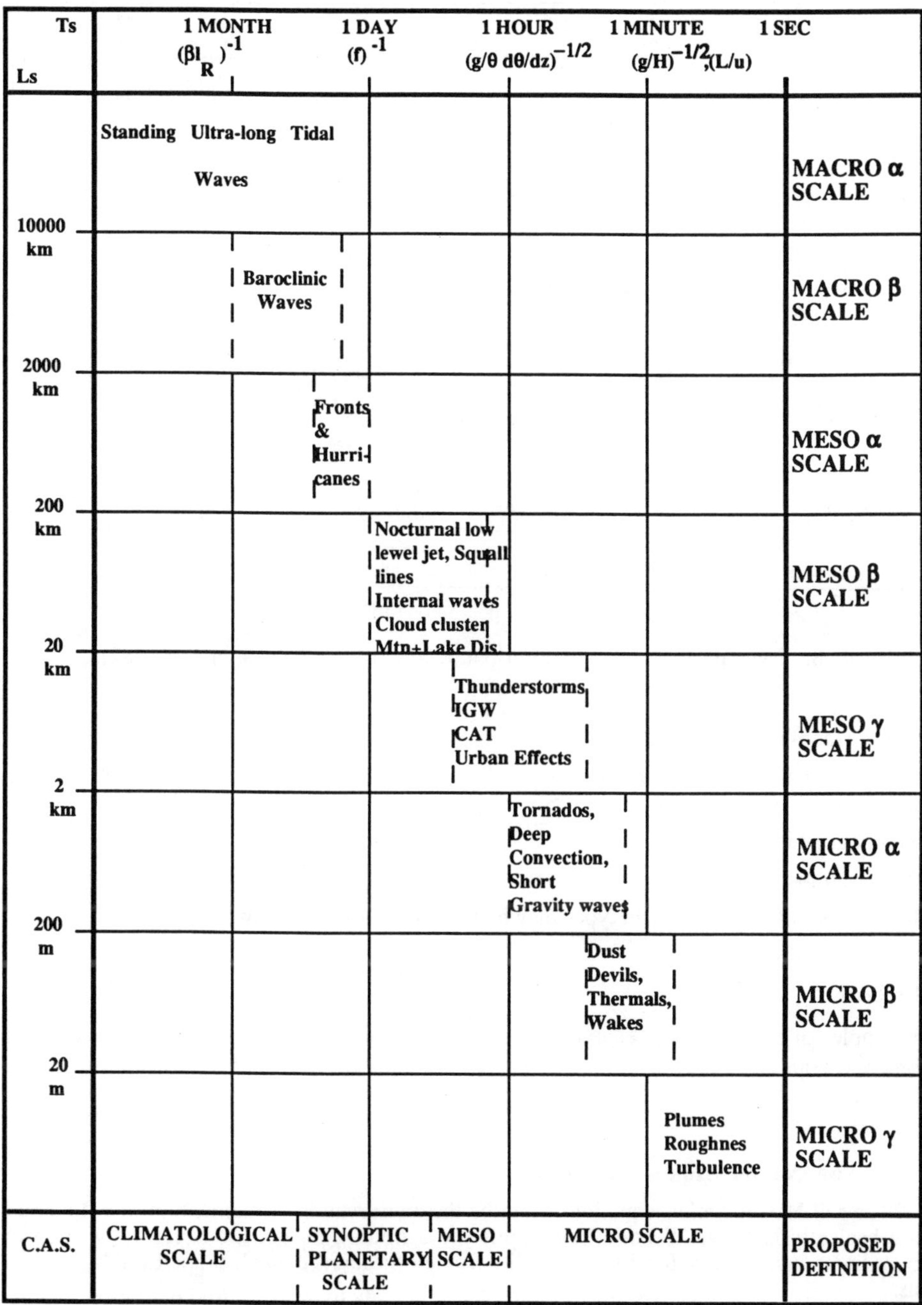

Figure 7. Scale definitions and different processes in the atmosphere with characteristic time and horizontal scales (after Orlanski 1975).

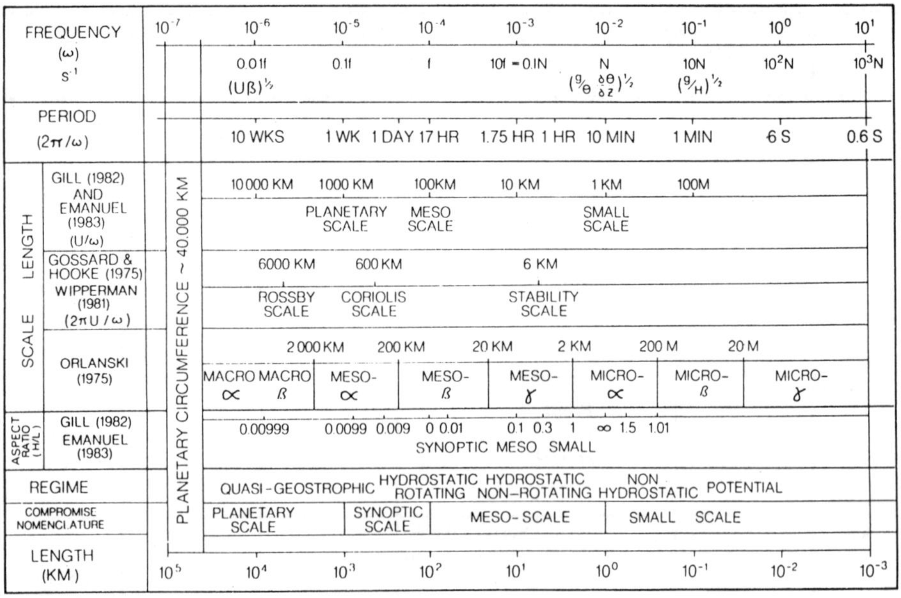

FREQUENCY (ω) s^{-1}		10^{-7} 10^{-6} 10^{-5} 10^{-4} 10^{-3} 10^{-2} 10^{-1} 10^{0} 10^{1} 0.01f ($(U\beta)^{1/2}$) 0.1f f 10f = 0.1N N ($(\frac{g}{\theta}\frac{\partial\theta}{\partial z})^{1/2}$) 10N ($(g/H)^{1/2}$) 10^2N 10^3N
PERIOD ($2\pi/\omega$)		10 WKS 1 WK 1 DAY 17 HR 1.75 HR 1 HR 10 MIN 1 MIN 6 S 0.6 S
SCALE LENGTH	GILL (1982) AND EMANUEL (1983) (U/ω)	10 000 KM 1000 KM 100KM 10 KM 1 KM 100M PLANETARY SCALE MESO SCALE SMALL SCALE
	GOSSARD & HOOKE (1975) WIPPERMAN (1981) ($2\pi U/\omega$)	6000 KM 600 KM 6 KM ROSSBY SCALE CORIOLIS SCALE STABILITY SCALE
	ORLANSKI (1975)	2 000 KM 200 KM 20 KM 2 KM 200 M 20 M MACRO α MACRO β | MESO- α | MESO- β | MESO- γ | MICRO- α | MICRO- β | MICRO- γ
ASPECT RATIO (H/L)	GILL (1982) EMANUEL (1983)	0.00999 0.0099 0.009 0 0.01 0.1 0.3 1 ∞ 1.5 1.01 SYNOPTIC MESO SMALL
REGIME		QUASI-GEOSTROPHIC HYDROSTATIC ROTATING HYDROSTATIC NON-ROTATING NON HYDROSTATIC POTENTIAL
COMPROMISE NOMENCLATURE		PLANETARY SCALE | SYNOPTIC SCALE | MESO-SCALE | SMALL SCALE
LENGTH (KM)		10^5 10^4 10^3 10^2 10^1 10^0 10^{-1} 10^{-2} 10^{-3}

PLANETARY CIRCUMFERENCE ~ 40.000 KM

Figure 8. Dynamical scalings of atmospheric motion (partially after Gill 1982).

1.3 Meso-scale circulations

Within the meso-scale identified above lies a host of circulations (Atkinson 1981). They may be sub-divided into two main types:

1)Topographically-induced

2)Free-atmosphere.

In turn each of these two main types may be further sub-divided in two sub-types:

a)Mechanically-forced

b)Thermally-forced

Examples of circulations in each type are: l(a) lee-waves, downslope winds; l(b) sea-land breezes, slope and valley breezes; 2(a) gravity waves; 2(b) severe convective storms, frontal circulations.

1.4 The Navier-Stokes Equations

The full Navier-Stokes equations describe the conservation of momentum in a fluid and are basically Newton's Second Law applied to a fluid. The equations can be written in coordinate independent form for a rotating frame of reference (such as the Earth's surface) as:

$$\frac{\partial \mathbf{V}}{\partial t} + \mathbf{V} \cdot \nabla \mathbf{V} = -\frac{1}{\rho} \nabla p - g\mathbf{k} - 2\Omega \times \mathbf{V} \tag{1.1}$$

where $\mathbf{V} = (u,v,w)$ is the velocity vector, t is time, ρ is air density, p is air pressure, g is the acceleration due to gravity and Ω the angular velocity of the system (so that $|\Omega| = \Omega$ is the angular speed of the Earth = $2\pi/24$ hours $\approx$ 7. 27 x 10^{-5} s^{-1}). Molecular viscosity is neglected here for simplicity; no generality is lost as the viscous diffusion terms will eventually be dominated by the turbulent Reynolds stresses, discussed later.

The full coriolis cross-product can be expanded to give terms in each of the components of the momentum equation (Hess 1959):

$2\Omega v \sin - 2\Omega w \cos\phi$ x- momentum

$-2\Omega u \sin\phi$ y- momentum

$2\Omega u \cos\phi$ z- momentum

which can be reduced to the familiar form with $f_c = 2\Omega \sin\phi$ as the coriolis parameter (which is ~10^{-4} s^{-1} for mid-latitude regions), where ϕ is latitude. This is possible because: in the x-momentum case the term in w is at least an order of magnitude less than the term in v; and in the z-momentum case the term as a whole is two orders of magnitude less than the other terms in the equation of vertical motion. The conservation of momentum equation can then be written in component form as:

$$\frac{\partial u}{\partial t} + u\frac{\partial u}{\partial x} + v\frac{\partial u}{\partial y} + w\frac{\partial u}{\partial z} = -\frac{1}{\rho}\frac{\partial p}{\partial x} + f_c v \tag{1.2a}$$

$$\frac{\partial v}{\partial t} + u\frac{\partial v}{\partial x} + v\frac{\partial v}{\partial y} + w\frac{\partial v}{\partial z} = -\frac{1}{\rho}\frac{\partial p}{\partial y} - f_c u \tag{1.2b}$$

$$\frac{\partial w}{\partial t} + u\frac{\partial w}{\partial x} + v\frac{\partial w}{\partial y} + w\frac{\partial w}{\partial z} = -\frac{1}{\rho}\frac{\partial p}{\partial z} - g \tag{1.2c}$$

To model atmospheric motions using these equations (1.2a)-(1.2c), a fourth equation needs to be satisfied in order for the model to conserve mass - the continuity equation:

$$\frac{\partial \rho}{\partial t} + \nabla \cdot (\rho \mathbf{V}) = 0 \tag{1.3}$$

which will be used later to modify the form of momentum equations (1.2a)-(1.2c).

1.5.1 Ensemble averaging

These four equations are valid for individual fluid elements, simply defined as groups of enough molecules so that their statistical (or 'group') properties are more important than their individual properties. The fluid motion predicted by the Navier-Stokes equations in their present form implicitly includes the 'mean' and turbulent flows (however the definition of these terms is chosen). The aim is to simulate the mean flow that is most likely to evolve from a given starting point without any turbulent fluctuations. To achieve this an averaging process, the ensemble average, is defined which separates out the mean and turbulent quantities. The properties of this average can be defined algebraically so that if ϕ is a variable associated with the 'real flow' (i.e. appears in one of Eqs. (1.2a)-(1.2c) above) and the mean and turbulent parts are written as $\overline{\phi}$ and ϕ'' respectively, the ensemble average of a quantity (or expression involving quantities) is denoted by an overbar and obeys the following rules:

(i) the average of a mean quantity is unchanged

$$\overline{(\overline{\phi})} = \overline{\phi} \tag{1.4a}$$

(ii) the average of a turbulent quantity is zero

$$\overline{\phi''} = 0 \tag{1.4b}$$

(iii) the averaging operation commutes with differentiation

$$\overline{\frac{\partial \phi''}{\partial t}} = \frac{\partial}{\partial t}(\overline{\phi''}) \tag{1.4c}$$

which, together with (ii) implies that $\overline{\frac{\partial \phi''}{\partial t}} = 0$ and similarly $\overline{\frac{\partial \phi''}{\partial x_i}} = 0$

1.5.2 Ensemble-averaging the momentum equations

In this section changes to the momentum equations under the averaging operation are examined. For simplicity only the u-component of the momentum is considered in detail; the other two components would receive identical treatment and the eventual results for all three components are presented at the end of the section.

First the flow variables are decomposed into mean-flow (denoted by an overbar) and turbulent (double-prime) parts.

$u = \overline{u} + u'', v = \overline{v} + v'', w = \overline{w} + w''$ for the velocity.

$\rho = \overline{\rho} + \rho''$ for the density (1.5)

p = p + p" for the pressure

Substituting this decomposition into Eq. (1.2a) results in

$$\frac{\partial(\overline{u}+u'')}{\partial t} + (\overline{u}+u'')\frac{\partial(\overline{u}+u'')}{\partial x} + (\overline{v}+v'')\frac{\partial(\overline{u}+u'')}{\partial y}$$
$$+(\overline{w}+w'')\frac{\partial(\overline{u}+u'')}{\partial z} = -(\overline{\rho}+\rho'')^{-1}\frac{\partial(\overline{p}+p'')}{\partial x} + f_c(\overline{v}+v'') \tag{1.6}$$

The next step is to ensemble average the transformed equation itself. Following the rules in Eqs. (1.4a)-(1.4c) above, the average of any mean quantity is unchanged, the average of a turbulent quantity is zero and hence, by omission, the average of products of turbulent quantities is non-zero: turbulence affects the mean motion through the correlation of two fluctuations as the mean of a single fluctuation is zero by definition. Averaging the transformed equation (1.6) and applying these rules, all of the terms involving turbulent fluctuations drop out except those involving means of products of turbulent quantities. Assuming the density fluctuation to be small, so that $\rho'' << \overline{\rho}$ (Assumption 1) and therefore $\frac{1}{(\overline{\rho}+\rho'')} \approx \frac{1}{\overline{\rho}}$, ensemble averaging Eq. (1.6) gives:

$$\frac{\partial \overline{u}}{\partial t} + \overline{u}_j\frac{\partial \overline{u}}{\partial x_j} = -\frac{1}{\overline{\rho}}\frac{\partial \overline{p}}{\partial x} + f_c\overline{v} - \overline{u''\frac{\partial u''}{\partial x}} - \overline{v''\frac{\partial u''}{\partial y}} - \overline{w''\frac{\partial u''}{\partial z}} \tag{1.7}$$

(1) (2) (3) (4) .... (5).............

where assumption 1 allows the neglect of ρ'' in term (3). The rules from Eqs. (1.4a)-(1.4c) reduce the first four terms of Eq. (1.6) to terms (1) and (2) of Eq. (1.7) except where the product of two turbulent fluctuations is averaged. In this case term (5) in Eq. (1.7), which gives the components of turbulent transfer, results from the advection terms in Eq. (1.6).

This procedure can be applied to each of the three momentum equations in turn to give (in tensor form):

$$\frac{\partial \bar{u}_i}{\partial t} + \bar{u}_j \frac{\partial \bar{u}_i}{\partial x_j} = -\frac{1}{\bar{\rho}} \frac{\partial \bar{p}}{\partial x_i} + f_c \varepsilon_{ij3} \bar{u}_j - g\delta_{i3} - \overline{u_j'' \frac{\partial u_i''}{\partial x_j}} \tag{1.8}$$

1.6 Approximations in the equations

We are now in a position to consider the approximations that may be made to the equations of momentum and continuity, the better to gain insight into their description of meso-scale airflows and also, in most cases, to simplify them. The geostrophic and hydrostatic approximations to the momentum equations are considered first. These are followed by the incompressible and anelastic approximations of the continuity equation and finally the Boussinesq approximation of the momentum equations is considered.

EQUATION OF HORIZONTAL MOTION

$$\frac{du}{dt} = \frac{\partial u}{\partial t} + u\frac{\partial u}{\partial x} + v\frac{\partial u}{\partial y} + w\frac{\partial u}{\partial z} = -\frac{1}{\rho}\frac{\partial p}{\partial x} + 2\Omega(v \sin\varnothing - w \cos\varnothing) + \frac{\partial}{\partial z}\left(K_z \frac{\partial u}{\partial z}\right) + \frac{\partial}{\partial x}\left(K_H \frac{\partial u}{\partial x}\right)$$

(1) (2) (3) (4) (5) (6) (7) (8) (9) (10)

(3)–(5): ADVECTIVE; (2)–(5): ACCELERATION; (6): PRESSURE GRADIENT FORCE; (7)–(8): CORIOLIS FORCE; (9): VERTICAL, (10): HORIZONTAL — TURBULENT DIFFUSION

LENGTH SCALE (L) (m)	ORDER ($m\ s^{-2}$)								
	$\frac{CU}{L}$	$\frac{U^2}{L}$	$\frac{U^2}{L}$	$\frac{U^2}{L}$	$\frac{\Delta P}{\rho L}$	fU	$\frac{fHU}{L}$	$\frac{KU}{H^2}$	$\frac{KU}{L^2}$
10^6	10^{-4}				10^{-3}	10^{-3}	10^{-5}	10^{-6}	10^{-10}
10^5	10^{-3}				10^{-2}	10^{-3}	10^{-4}	10^{-6}	10^{-8}
10^4	10^{-2}				10^{-1}	10^{-3}	10^{-3}	10^{-6}	10^{-6}
10^3	10^{-1}				10^{0}	10^{-3}	10^{-2}	10^{-6}	10^{-4}
10^2	10^{0}				10^{1}	10^{-3}	10^{-1}	10^{-6}	10^{-2}

Figure 9. The equation of horizontal motion applicable to the atmosphere. Characteristic horizontal length scale is on the left.

1.6.1 Geostrophic approximation

The magnitudes of the terms in the equations of horizontal motion (Eq 1.8) vary over several orders of magnitude. Figure 9 illustrates the u-momentum equation (in which the form of the turbulent transfer terms is explained in section 11) and shows the terms scaled by using basic elements of the flow mechanism. For airflows with horizontal length scales of hundreds or more kilometres (the length scale is on the left) the horizontal pressure gradient force term and the coriolis term dominate the equation, to an extent that the equation can be reduced to only those two terms with a high degree of validity, giving the form for the *u*-momentum equation of

$$\frac{1}{\rho}\frac{\partial p}{\partial x} = f_c v \tag{1.9}$$

A similar form exists for the *v*-momentum equation. Eq (1.9) describes a state known as geostrophic balance. The velocity *v* is known as the geostrophic wind. This concept is widely used in large-scale meteorology generally and in weather forecasting in particular. At the smaller length scales found in meso-scale meteorology the other terms in the momentum equations cannot be validly ignored so the geostrophic approximation plays a minor role in the analysis of meso-scale airflows.

EQUATION OF VERTICAL MOTION

$$\frac{dw}{dt} = \frac{\partial w}{\partial t} + u\frac{\partial w}{\partial x} + v\frac{\partial w}{\partial y} + w\frac{\partial w}{\partial z} = -\frac{1}{\rho}\frac{\partial p}{\partial z} - g + 2\Omega u \cos\varnothing + \frac{\partial}{\partial z}\left(K_Z\frac{\partial w}{\partial z}\right) + \frac{\partial}{\partial x}\left(K_H\frac{\partial w}{\partial x}\right) + \frac{\partial}{\partial y}\left(K_H\frac{\partial w}{\partial y}\right)$$

(1) (2) (3) (4) (5) (6) (7) (8) (9) (10) (11)

(2)–(5): ACCELERATION, of which (3)–(5): ADVECTIVE; (6): PRESSURE GRADIENT FORCE; (7): GRAVITY; (8): CORIOLIS FORCE; (9): VERTICAL, (10)–(11): HORIZONTAL — TURBULENT DIFFUSION

LENGTH SCALE (L) (m)	ORDER (m s-2)									
	$\frac{U^2H}{L^2}$	$\frac{U^2H}{L^2}$	$\frac{U^2H}{L^2}$	$\frac{U^2H}{L^2}$	$\frac{\Delta P}{\rho H}$	10^1	fU	$\frac{KU}{HL}$	$\frac{KUH}{L^3}$	$\frac{KUH}{L^3}$
10^6	10^{-6}				10^1	10^1	10^{-3}	10^{-8}	10^{-12}	
10^5	10^{-4}				10^1	10^1	10^{-3}	10^{-7}	10^{-9}	
10^4	10^{-2}				10^1	10^1	10^{-3}	10^{-6}	10^{-6}	
10^3	10^0				10^1	10^1	10^{-3}	10^{-5}	10^{-3}	
10^2	10^2				10^1	10^1	10^{-3}	10^{-4}	10^0	

Figure 10. The equation of vertical motion applicable to the atmosphere. Characteristic horizontal length scale is on the left.

1.6.2 Hydrostatic approximation

Figure 10 shows the equation of vertical motion (Eq 1.2c) which is scaled in the same way as Figure 9. Inspection of Figure 10 reveals that the vertical pressure gradient and gravity terms are several orders of magnitude greater than the other terms when applied to airflows with horizontal length scales of hundreds or more kilometres. This leads to the hydrostatic approximation where the equation of vertical motion is reduced to a balance of those two terms, thus

$$-\frac{1}{\rho}\frac{\partial p}{\partial z} = g \tag{1.10}$$

This approximation is also a mainstay of large scale meteorology. Its use in meso-scale meteorology was widespread until the late 1970s and indeed it is still has a role to-day . Pielke's (1984) book on modelling meso-scale motions uses the hydrostatic approximation throughout, arguing (after Martin and Pielke (1983)) that the approximation is valid with length scales as small as a few tens of kilometres. Over the past twenty years non-hydrostatic models have become more commonplace (Miller and White 1984): one of the first was that by Tapp and White (1976). In the remainder of this chapter the non-hydrostatic state will be used.

1.6.3 Incompressible and Anelastic approximations

It is possible to gain further insight into the nature of meso-scale airflows by examining the continuity equation (Eq 1.3). This allows the identification of compressible and incompressible flows and the anelastic characteristic (see Ogura and Phillips (1962) and Dutton and Fichtl (1969)). The specific volume (α), defined as $\alpha=1/\rho$, is decomposed further by writing the grid average ($\overline{\alpha}$) in terms of a synoptic scale background density (α_0), which is defined as the horizontal average of $\overline{\alpha}$ across the computational domain, so that α_0 depends, at most, only on z. Then the mesoscale deviation from the synoptic scale is defined as $\alpha' = \overline{\alpha} - \alpha_0$, which is the only part of the density that is allowed to vary significantly in time and space. Using this decomposition the continuity equation (Eq 1.3) can be written as:

$$\frac{\partial(\alpha_0+\alpha')}{\partial t}+u_j\frac{\partial}{\partial x_j}(\alpha_0+\alpha')=(\alpha_0+\alpha')\frac{\partial u_j}{\partial x_j} \tag{1.11}$$

Application of scale analysis demonstrates that all terms of this equation involving the perturbation density α' can be neglected in favour of the terms involving the background density α_0 if $|\alpha'/\alpha_0| << 1$ (Assumption 2), i.e. the mesoscale density deviation is small compared to the background reference density profile. Even the significance of the time variation of α' can be reduced to this condition: it is satisfied to better than 5% in all mesoscale situations. Using this fact the $\frac{\partial \alpha'}{\partial t}$ and $\frac{\partial \alpha'}{\partial x_j}$ terms in Eq. (1.11) can be neglected. Also assumption 2 allows the bracket on the right hand side of Eq. (1.11) to be re-written as $(\alpha_0+\alpha')=\alpha_0\times\left(1+\alpha'/\alpha_0\right)\approx\alpha_0$ so Eq (1.11) reduces to:

$$\underset{(1)}{\frac{\partial \alpha_0}{\partial t}}+\underset{(2)}{\alpha_0\frac{\partial u_j}{\partial x_j}}=\underset{(3)}{u_j\frac{\partial \alpha_0}{\partial x_j}} \tag{1.12}$$

From the definition of the mesoscale background density, α_0, as a horizontal average over the mesoscale domain, α_0 is at most a function of z alone; no horizontal- or time-dependence is allowed, and therefore in Eq. (1.12), term (1) is zero, as is term (3) for j=1 or 2 (but not necessarily for j=3). Thus the continuity equation is reduced to:

$$\alpha_0\frac{\partial u_j}{\partial x_j}=w\frac{\partial \alpha_0}{\partial z} \tag{1.13}$$

Eq (1.13) can be re-written as

$$w^{-1}\frac{\partial u_j}{\partial x_j}=\alpha_0^{-1}\frac{\partial \alpha_0}{\partial z} \tag{1.14}$$

The term on the left hand side of Eq (1.14) can be considered to be the inverse of the vertical length scale of the air circulation of interest (L_z). A length scale for the vertical variation of background density, the density scale depth of the atmosphere, can be defined as $H_a^{-1} \equiv \frac{1}{\alpha_0}\frac{d\alpha_0}{dz}$ which is ~ 8 km in the troposphere. Hence these length scales measure the scales of the left hand and right hand sides respectively of Eq (1.14). This can be used to identify the incompressible and anelastic forms of the continuity equation.

1.6.3.1 Incompressibility. If the vertical length scale of a circulation is much less than the density scale depth of the atmosphere, ie $L_z << H_a$, so that the left hand side of Eq (1.14) dominates the right hand side, the atmosphere can be treated as incompressible and this state is described by the expression

$$\frac{\partial u_j}{\partial x_j} = 0 \tag{1.15a}$$

This is known as either the incompressible continuity equation because there is no variation in density (ρ), it being absent from the equation; or the shallow continuity equation, because of the relative magnitude of L_z.

1.6.3.2 Anelasticity. If the vertical length scale of the circulation is of the same order as the scale depth of the atmosphere, ie $L_z \sim H_a$, neither side of Eq (1.14) may justifiably be ignored. It is convenient to add to the right hand side of Eq (1.13) the (zero) horizontal gradient terms for α_0 so that the equation can be re-written as

$$0 = \alpha_0 \frac{\partial u_j}{\partial x_j} - w\frac{\partial \alpha_0}{\partial z} = \alpha_0 \frac{\partial u_j}{\partial x_j} - u_j\frac{\partial \alpha_0}{\partial x_j} = \frac{\partial(\alpha_0^{-1} u_j)}{\partial x_j}$$

and identifying α_0 with $1/\rho_0$ results in:

$$\frac{\partial(\rho_0 u_j)}{\partial x_j} = 0 \tag{1.15b}$$

Eq (1.15b) is known as either the anelastic form of the continuity equation, because it prevents the formation of sound waves in models; or the deep form of the continuity equation, because of the relative magnitude of L_z.

Meso-scale flows are frequently shallow with respect to H_a, so the incompressible equation is acceptable and has been used in many situations, especially analytical models where its simplicity is useful. However the anelastic equation allows the background density to vary in the vertical (whereas for the incompressible state it must be constant throughout) and as such is more general without being much more complicated. It has been the preferred form in the years since its clarification.

1.7. The turbulent transfer terms

Whichever form of the continuity equation is adopted, it allows the momentum equation to be rewritten so that the turbulent terms take the form of pure divergences, Using the anelastic form (the same analysis applies to the incompressible) the components of velocity in Eq. (1.15b) can be expressed as the sum of mean and turbulent parts (using Eq. (1.5)):

$$\frac{\partial}{\partial x_j}\left[\rho_0(\bar{u}_j + u_j'')\right] = 0 \tag{1.16}$$

Then averaging Eq. (1.16) using the rules contained in Eqs. (1.4a)-(1.4c), the continuity equation for the mean flow becomes $\frac{\partial}{\partial x_j}(\rho_0 \bar{u}_j) = 0$. Subtracting this equation from Eq. (1.15b) gives $\frac{\partial}{\partial x_j}(\rho_0 u_j'') = 0$. Using this the momentum equation (Eq(1.7)) can be re-written by taking the derivative in the turbulent flux terms outside of the average (denoted by an overbar), which puts these terms in their familiar flux divergence form

$$\overline{u_j'' \frac{\partial u_i''}{\partial x_j}} = \frac{1}{\rho_0}\left[\overline{\rho_0 u_j'' \frac{\partial u_i''}{\partial x_j}}\right] = \frac{1}{\rho_0}\left[\overline{\frac{\partial(\rho_0 u_j'')u_i''}{\partial x_j}}\right] \quad \text{since} \quad \frac{\partial(\rho_0 u_j'')}{\partial x_j} = 0$$

$$= \frac{1}{\rho_0}\left[\frac{\partial \rho_0 \overline{u_i'' u_j''}}{\partial x_j}\right]$$

Using this in Eq. (1.7) and the similarly transformed v- and w-momentum equations the Navier-Stokes equations can be re-written as:

$$\frac{\partial \bar{u}}{\partial t} + \bar{u}\frac{\partial \bar{u}}{\partial x} + \bar{v}\frac{\partial \bar{u}}{\partial y} + \bar{w}\frac{\partial \bar{u}}{\partial z} = -\frac{1}{\bar{\rho}}\frac{\partial \bar{p}}{\partial x} + f_c \bar{v} - \frac{1}{\rho_0}\left(\frac{\partial}{\partial x}\rho_0 \overline{u''u''} + \frac{\partial}{\partial y}\rho_0 \overline{v''u''} + \frac{\partial}{\partial z}\rho_0 \overline{w''u''}\right) \tag{1.17a}$$

$$\frac{\partial \bar{v}}{\partial t} + \bar{u}\frac{\partial \bar{v}}{\partial x} + \bar{v}\frac{\partial \bar{v}}{\partial y} + \bar{w}\frac{\partial \bar{v}}{\partial z} = -\frac{1}{\bar{\rho}}\frac{\partial \bar{p}}{\partial y} - f_c \bar{u} - \frac{1}{\rho_0}\left(\frac{\partial}{\partial x}\rho_0 \overline{u''v''} + \frac{\partial}{\partial y}\rho_0 \overline{v''v''} + \frac{\partial}{\partial z}\rho_0 \overline{w''v''}\right) \tag{1.17b}$$

$$\frac{\partial \bar{w}}{\partial t} + \bar{u}\frac{\partial \bar{w}}{\partial x} + \bar{v}\frac{\partial \bar{w}}{\partial y} + \bar{w}\frac{\partial \bar{w}}{\partial z} = -\frac{1}{\bar{\rho}}\frac{\partial \bar{p}}{\partial z} - g - \frac{1}{\rho_0}\left(\frac{\partial}{\partial x}\rho_0 \overline{u''w''} + \frac{\partial}{\partial y}\rho_0 \overline{v''w''} + \frac{\partial}{\partial z}\rho_0 \overline{w''w''}\right) \tag{1.17c}$$

which include a widely used form of sub-grid scale turbulent flux terms (double primed). Two species of density are present - ρ_0 and $\bar{\rho}$. The ρ_0 in the turbulent transport terms arises from the anelastic equation and would not be present with the incompressible assumption. The $\bar{\rho}$ in each pressure-gradient term can be beneficially replaced by a ρ_0 and this is dealt with in the following section.

1.8. The Boussinesq approximation

A further approximation - the Boussinesq approximation (Boussinesq (1903), Gray and Giorgini (1976), Spiegal and Veronis (1960)) - is commonly applied to the momentum equations. This simplifies the model equations by replacing all reference to density with the constant background field. Here the procedure is divided into three parts for clarity:

1.8.1 The hydrostatic part

The pressure gradient term in Eq. (1.17c), and only the term from the w -momentum equation, contains a large, hydrostatic part which is partially balanced by gravity. In other words, the pressure gradient and gravitational acceleration partially balance each other and it is sensible to remove these unchanging parts from the equations. This is achieved by assuming that the background state is in hydrostatic balance so that the synoptic state variables obey the following equation:

$$\frac{\partial p_0}{\partial z} = -\rho_0 g \quad \text{or} \quad \alpha_0 \frac{\partial p_0}{\partial z} = -g \tag{1.18}$$

Then re-writing the pressure gradient and gravitational acceleration terms in Eq. (1.17c) using the decomposition of $\overline{\alpha}$ into $\alpha_0 + \alpha'$ and similarly for $\overline{p}$

$$\overline{\alpha}\frac{\partial \overline{p}}{\partial z} + g = (\alpha_0 + \alpha')\frac{\partial}{\partial z}(p_0 + p') + g$$

$$= \alpha_0\left(1 + \alpha'/\alpha_0\right)\frac{\partial p'}{\partial z} + \alpha_0\left(1 + \alpha'/\alpha_0\right)\frac{\partial p_0}{\partial z} + g$$

Using the hydrostatic relation (Eq. (1.18)) the right hand side becomes

$$= \alpha_0\left(1 + \alpha'/\alpha_0\right)\frac{\partial p'}{\partial z} - \left(1 + \alpha'/\alpha_0\right)g + g$$

so that the background pressure-gradient cancels with g to leave

$$= \alpha_0\left(1 + \alpha'/\alpha_0\right)\frac{\partial p'}{\partial z} - \frac{\alpha'}{\alpha_0}g$$

Assumption 2 allows the second term in parentheses to be neglected so that $\alpha_0\left(1 + \alpha'/\alpha_0\right) \approx \alpha_0$ giving $-\alpha_0\frac{\partial p'}{\partial z} + \frac{\alpha'}{\alpha_0}g$ as the approximation for $-\overline{\alpha}\frac{\partial \overline{p}}{\partial z} - g$ in the w-momentum equation.

1.8.2 The gravitational acceleration

The dependence of the gravitational acceleration term, $g(\alpha'/\alpha_0)$, on the mesoscale density perturbation, α', can be removed by logarithmically differentiating the ideal gas equation $\alpha\text{p} = \text{RT}$ and using the potential temperature defined as $\theta = T\left(\frac{1000mb}{p}\right)^{R/c_p}$. Substituting the mesoscale perturbation and synoptic state variables the following ensues for deep circulations where $L_z \sim H_a$:

$$\frac{\alpha'}{\alpha_0} \approx \frac{\theta'}{\theta_0} - \frac{c_v}{c_p}\frac{p'}{p_0} \tag{1.19a}$$

and using a similar scale analysis to that used for the continuity equation this can in turn be approximated for shallow circulations ($L_z << H_a$)as:

$$\frac{\alpha'}{\alpha_0} \approx \frac{\theta'}{\theta_0} \tag{1.19b}$$

Thus α'/α_0 can be re-written in terms of the potential temperature so that, assuming in this case that the form appropriate for $L_z << H_a$ is used, the w-momentum equation becomes:

$$\frac{\partial \overline{w}}{\partial t}+\overline{u}\frac{\partial \overline{w}}{\partial x}+\overline{v}\frac{\partial \overline{w}}{\partial y}+\overline{w}\frac{\partial \overline{w}}{\partial z}=-\frac{1}{\rho_0}\frac{\partial p'}{\partial z}+\frac{\theta'}{\theta_0}g$$
$$-\frac{1}{\rho_0}\left(\frac{\partial}{\partial x}\rho_0\overline{u''w''}+\frac{\partial}{\partial y}\rho_0\overline{v''w''}+\frac{\partial}{\partial z}\rho_0\overline{w''w''}\right) \tag{1.20}$$

The latter form for the gravitational acceleration term (given by Eq. (1.19b)) is often used along with the anelastic equation even though Eq. (1.19b) relies on a more stringent condition than the anelastic approximation to the continuity equation.

1.8.3 The mesoscale pressure perturbation p'

Turning attention to the u- and v-momentum equations (1.17a) and (1.17b), the pressure field can be decomposed into its mesoscale perturbation and background state (as defined previously while simplifying the continuity equation) so that $\overline{p}=p_0+p'$. Since the background pressure is assumed hydrostatic and uniform in x and y, p_o is a function of z alone so that:

$$\frac{\partial \overline{p}}{\partial x}=\frac{\partial}{\partial x}(p_0+p')=\frac{\partial p'}{\partial x}$$

and similarly for the y--derivative. Thus the derivatives of $\overline{p}$ in the u- and v-momentum equations are replaced by derivatives of p' and thus $\overline{p}$ is removed from all the momentum equations.

The grid-averaged density $\overline{\rho}$ still remains in the u- and v--momentum equations (in the pressure gradient terms): it is desirable to remove this, if possible, as it is a further unknown. Assumption 2 (for its specific-volume counterpart α') allows us to neglect α' in favour of α_0, so writing $\frac{1}{\overline{\rho}}$ as $\overline{\alpha}=\alpha_0+\alpha'$ which equals $\alpha_0\times\left(1+\frac{\alpha'}{\alpha_0}\right)$ and applying assumption 2 implies that this is $\approx \alpha_0$ which in turn equals $1/\rho_0$, so that all occurrences of $\overline{\rho}$ can be replaced by ρ_0 as desired.

Using the results from the three steps outlined above the momentum equations are written in the Boussinesq form, using the simplest form for the gravitational acceleration term, from Sec 8.2 above, as follows:

$$\frac{\partial \overline{u}}{\partial t}+\overline{u}\frac{\partial \overline{u}}{\partial x}+\overline{v}\frac{\partial \overline{u}}{\partial y}+\overline{w}\frac{\partial \overline{u}}{\partial z}=-\frac{1}{\rho_0}\frac{\partial p'}{\partial x}+f_c\overline{v}$$
$$-\frac{1}{\rho_0}\left(\frac{\partial}{\partial x}\rho_0\overline{u''u''}+\frac{\partial}{\partial y}\rho_0\overline{v''u''}+\frac{\partial}{\partial z}\rho_0\overline{w''u''}\right) \tag{1.21a}$$

$$\frac{\partial \overline{v}}{\partial t}+\overline{u}\frac{\partial \overline{v}}{\partial x}+\overline{v}\frac{\partial \overline{v}}{\partial y}+\overline{w}\frac{\partial \overline{v}}{\partial z}=-\frac{1}{\rho_0}\frac{\partial p'}{\partial y}-f_c\overline{u}$$
$$-\frac{1}{\rho_0}\left(\frac{\partial}{\partial x}\rho_0\overline{u''v''}+\frac{\partial}{\partial y}\rho_0\overline{v''v''}+\frac{\partial}{\partial z}\rho_0\overline{w''v''}\right) \tag{1.21b}$$

$$\frac{\partial \overline{w}}{\partial t}+\overline{u}\frac{\partial \overline{w}}{\partial x}+\overline{v}\frac{\partial \overline{w}}{\partial y}+\overline{w}\frac{\partial \overline{w}}{\partial z}=-\frac{1}{\rho_0}\frac{\partial p'}{\partial z}+\frac{\theta'}{\theta_0}g$$
$$-\frac{1}{\rho_0}\left(\frac{\partial}{\partial x}\rho_0\overline{u''w''}+\frac{\partial}{\partial y}\rho_0\overline{v''w''}+\frac{\partial}{\partial z}\rho_0\overline{w''w''}\right) \quad (1.21c)$$

The appearance of ρ_0 in the turbulent flux divergence terms is due, as was noted in Sec 7 to the use of the more general anelastic equation rather then the simpler incompressible form . Since the anelastic equation allows ρ_0 to vary in the vertical then ρ_0 cannot be taken out from under the *z*-derivative term (in the w''- fluxes) without generating an extra term of the form $\overline{w''u''}\frac{\partial \rho_0}{\partial z}$ which breaks the symmetry of the equations.

1.9. Flux form of the momentum equations

The advection terms of the momentum equations can be written in flux-form using either the anelastic or incompressible continuity equations and this is commonly done to simplify the finite-difference form of the fluid equations in numerical models.

Assuming the anelastic condition (Eq 1.15b) for generality the advection term $u_j\frac{\partial u_i}{\partial x_j}$ can be re-written in terms of the momentum flux $(\rho_0 u_i u_j)$. Multiplying the advection term by ρ_0/ρ_0 allows the term to be re-written as $\frac{1}{\rho_0}\left(\rho_0 u_j\frac{\partial u_i}{\partial x_j}\right)$. If the $\rho_0 u_j$ term is moved inside the derivative the following relation is true:

$$\frac{1}{\rho_0}\frac{\partial}{\partial x_j}(\rho_0 u_i u_j)=\frac{1}{\rho_0}\left(\rho_0 u_j\frac{\partial u_i}{\partial x_j}+u_i\frac{\partial \rho_0 u_j}{\partial x_j}\right)$$
$$=u_j\frac{\partial u_i}{\partial x_j}$$

since the continuity equation states that $\frac{\partial}{\partial x_j}(\rho_0 u_j)=0$. Thus the advection term $\left(u_j\frac{\partial u_i}{\partial x_j}\right)$ is re-written as $\frac{1}{\rho_0}\frac{\partial}{\partial x_j}(\rho_0 u_i u_j)$ which is known as the flux-form of that term. Consequently the convective derivative on the left hand side of Eqs (1.21a)-(1.21c) becomes $\frac{\partial u_i}{\partial t}+\frac{1}{\rho_0}\frac{\partial}{\partial x_j}(\rho_0 u_i u_j)$.

1.10 Conservation of a scalar quantity

The Navier-Stokes equations describe the conservation of a vector quantity, the momentum, and the continuity equation describes the conservation of mass. There is one more flow property that needs to be catered for - that of the conservation of a scalar property that is associated with the fluid element, such as temperature, water vapour or a pollutant. In the same way as for the momentum equation, the conservation of a scalar involves the time-rate-of-change of the quantity balanced by any sources and sinks of the scalar. This is illustrated by considering the conservation of heat, which is represented by the equation of potential temperature. The conservation of heat equation can be written as:

$$\frac{\partial\overline{\theta}}{\partial t}+\overline{u}\frac{\partial\overline{\theta}}{\partial x}+\overline{v}\frac{\partial\overline{\theta}}{\partial y}+\overline{w}\frac{\partial\overline{\theta}}{\partial z}=\overline{S}_\theta$$
$$-\frac{1}{\rho_0}\left(\frac{\partial}{\partial x}\rho_0\overline{u''\theta''}+\frac{\partial}{\partial y}\rho_0\overline{v''\theta''}+\frac{\partial}{\partial z}\rho_0\overline{w''\theta''}\right) \tag{1.22}$$

and the form of the conservation equation for any other scalar is the same, with the temperature replaced by the other scalar . The source term $\overline{S}_\theta$ contains all the contributions due to diabatic effects (for example, long-wave radiation may be an important effect in nocturnal flows) and its specification can be a complicated procedure. The turbulent flux divergence terms arise from the averaging but a scalar such as heat does not necessarily mix in the same way as the momentum in Eqs. (1.21a)-(1.21c). In practice the processes of the turbulent flux divergences for heat, water and other scalar quantities are generally assumed to be the same.

1.11 Turbulence closure

The turbulence flux divergence terms $\frac{\partial\overline{u_i''u_j''}}{\partial x_j}$ and $\frac{\partial\overline{\theta''u_j''}}{\partial x_j}$ arise from ensemble-averaging the Navier-Stokes equations. Their specification in terms of other variables is called the closure problem because the number of unknowns in the set of equations for turbulent flow is larger than the number of equations (Stull (1988 Ch. 6) gives a good description of this problem). It is possible to derive equations for the unknown turbulent fluxes $\left(\overline{u_i''u_j''}\right)$, which are second order terms as they involve products of two turbulent quantities, but these equations inevitably involve third order terms (products of three turbulent quantities), and so-on. As the order of the turbulent terms in the prognostic equations increases so does the number of new unknowns.

To overcome this problem the unknown turbulent quantities are approximated in terms of known variables and various parameters. These parameters must be determined outside the model and are usually derived either from observational data or more detailed models.The level of closure refers to the order of the products of turbulent quantities that are parameterised in terms of other variables: for example a first order closure specifies the terms involving $\left(\overline{u_i''u_j''}\right)$ and $\left(\overline{\theta''u_j''}\right)$ directly in terms of the mean variables $\overline{u}_i$ and $\overline{\theta}$, whereas a second order closure expresses the third moment correlations (terms of the form $\left(\overline{u_i''u_j''u_k''}\right)$) using terms up to second order, and so-on.

The two types of closure used in numerical models are called local and non-local. Local closures express turbulent correlation terms at a point using values from around that point (for example, not more than two points away using standard, second-order finite-differences), whereas non-local closures use values away from the point in question. The non-local approach can be better than a local closure because it acknowledges that large turbulent eddies can mix over large distances, but it is more difficult to apply as data from many points are required. The most common form of local closure replaces the turbulent flux divergences with diffusive terms and is known as K-theory or gradient transport theory (Dyer 1974) so that for an element χ

$$\frac{\partial}{\partial x_j}(\rho_0\overline{u_j''\chi''})=\frac{\partial}{\partial x_j}\left(-\rho_0 K\frac{\partial\overline{\chi}}{\partial x_j}\right) \tag{1.23}$$

which is analogous to the way molecular viscosity appears in the Navier-Stokes equations if written in terms of the Reynolds stress:

$\tau_{ij} = \rho_0 \nu \frac{\partial \bar{u}_i}{\partial x_j}$ the stress due to molecular viscosity and

$\tau_{ij} = \rho_0 K_m \frac{\partial \bar{u}_i}{\partial x_j}$ the Reynolds stress.

Here the kinematic viscosity ν is analogous to the eddy viscosity (diffusivity or exchange coefficient) for momentum, K_m and the product ρK_m is called the Austausch coefficient. This is known as gradient transport theory since, for positive diffusivities, the turbulent flux flows down the local gradient of the mean variable (from high values to low values).

One assumption often made concerning the K's is that the vertical turbulent flux divergences are much larger than their horizontal equivalentst so the latter are neglected with respect to the former. Over horizontal terrain in weakly-stable conditions this may be a satisfactory approximation, but stable conditions inhibit vertical motions so the turbulent eddies are squashed in the vertical and this increases the importance of the horizontal mixing.

The diffusivities of momentum, heat and water may differ in both specification and magnitude, but there is little agreement as to the exact nature of these differences. It is widespread practice to assume that the magnitudes of vertical diffusivities are equal. In the following the case of diffusion of momentum is used to illustrate the different formulations of vertical diffusivities in general, of which there is an abundance (McBean (1979)). Horizontal diffusivities are frequently ignored in numerical models or at best are simply specified as constants. The formulation of vertical diffusivities in models is aided by dividing the domain into two layers which fairly closely reflect the real structure of the planetary boundary layer, the area of most concern here. First is the surface layer which lies in the bottom few metres of the atmosphere. In that layer Monin-Obukhov similarity theory is deemed to hold. Secondly, the Ekman layer overlies the surface layer and extends to the top of the planetary boundary layer (c 1km). The vertical diffusivity is established in different ways in the two layers.

In the surface layer the use of Monin-Obukhov theory (Monin and Obukhov 1954) means that it is possible to eliminate explicit numerical calculations of the wind, temperature and humidity profiles in that layer. The theory is stated in the form of a dimensionless vertical profiles, exemplified here by wind shear (u-component used for illustration) and lapse rate, which are respectively

$$\frac{k_0 z}{u_*} \frac{\partial u}{\partial z} = F_m(z/L) \tag{1.24}$$

$$\frac{k_0 z}{\theta_*} \frac{\partial \theta}{\partial z} = F_h(z/L) \tag{1.25}$$

where $u_* = (\tau/\rho)^{1/2}$ is the friction velocity, $\tau = -\rho\overline{u'w'}$ is the shear stress, $\theta_* = -(\overline{w'\theta'})/u_*$ is the temperature scale, k_0 is von Karman's constant and z is height. F_m and F_h are the non-dimensional, universal functions of the stability parameter z/L which have to be determined from observations. They are frequently assumed to take the same form, but the validity of this assumption is open to question. The most accepted form of the functions, illustrated by that for wind, is

$$F_m(z/L) = [1 - 15(z/L)]^{-1/4} \qquad \frac{z}{L} < 0 \tag{1.26}$$

$$F_m(z/L) = 1 + 4.7(z/L) \qquad \frac{z}{L} > 0 \tag{1.27}$$

where the Monin-Obukhov length L is

$$L = -u_*^3 / \left[k_0 (g/\theta) \overline{w''\theta''} \right] \tag{1.28}$$

where g is the acceleration of gravity (Businger et al (1971), Dyer and Hicks (1970)). These functions can readily be integrated in Eqs (1.24) and (1.25) thus allowing the profiles of u and θ to be established throughout the surface layer (Paulson 1970). The eddy diffusivities for momentum and heat in the surface layer are given respectively by

$$K_m = u_* k_0 z / F_m(z/L) \tag{1.29}$$

$$K_h = u_* k_0 z / F_h(z/L) \tag{1.30}$$

In the Ekman layer two main types of first order schemes are available. First is the geometrical specification such as the widely used version suggested by O'Brien (1970). This takes the form

$$K(z) = K_a + \left[(z_a - z)^2 / (\Delta z)^2 \right] \times \left\{ K_b - K_a + (z - z_b) \left[K_b' + 2(K_b - K_a)/\Delta z \right] \right\} \tag{1.31}$$

where z is height, z_a and z_b are the heights of the top of the Ekman and surface layers respectively, K_a and K_b are the diffusivties at heights z_a and z_b respectively, Δz is the height difference between z_a and z_b and $K_b' \equiv \partial K/\partial z$ at height z_b. This form is perhaps the most sophisticated of a type of specification that has a pedigree stretching back to the early years of this century when the vertical diffusivities were simply specified as being constant in both space and time. The second form is based on a mixing length approach. A typical form is that originally suggested by Blackadar (1962) and generalised for stability effects by Estoque and Bhumralkar (1969) and Yu (1977) . This takes the form

$$K_m = l^2 \left[\left(\frac{\partial u}{\partial z} \right)^2 + \left(\frac{\partial v}{\partial z} \right)^2 \right]^{1/2} G \tag{1.32}$$

where the mixing length l is given by

$$l = \frac{k_0 (z + z_0)}{1 + k_0 \left(\dfrac{z + z_0}{\lambda} \right)} \qquad \text{where} \qquad \lambda = \frac{27 \mathbf{V}_g \times 10^{-5}}{f_c}$$

z is the height, z_0 is the roughness length, $\mathbf{V}_g$ is the geostrophic wind and f_c is the coriolis parameter. G is a stability function

$$G = (1 + \beta \mathrm{Ri})^{-2} \qquad \mathrm{Ri} \geq 0$$

$$G = (1 - \beta \mathrm{Ri})^{2} \qquad \mathrm{Ri} < 0$$

where β is a constant, usually given the value 3. The Richardson number is given by

$$\mathrm{Ri} \equiv \frac{g \, \partial\theta/\partial z}{\theta (\partial \mathbf{V}/\partial z)^2}$$

Higher order closure schemes (Mellor and Yamada 1974) are typified by those involving the Turbulent Kinetic Energy (TKE), which is defined as $E = 0.5(\overline{u''^2} + \overline{v''^2} + \overline{w''^2})$. The diffusivity is expressed as a function of this TKE

$K = f(E)$

which thus requires evaluation of E. This is done with a prognostic equation for E of the following form

$$\frac{\partial E}{\partial t} = \text{f(Production, Transfer, Dissipation of TKE)} \qquad (1.33)$$

The terms on the right hand side of Eq (1.33) include higher order terms to ensure closure. In running a numerical model that includes this formulation, Eq (1.33) is solved at each time step to give values of E which are then used to calculate K. It is possible to construct closures of order higher than two but these are complex and expensive in computer time.

1.12 Flow over non-flat orography

Most meso-scale models prior to the 1970s were for airflows over flat terrain. The equations presented throughout this chapter are suitable for such modelling. Most flows actually occur, of course, over non-uniform orography and consequently in numerical models of such flows it is necessary to transform the coordinates so that the equations accurately represent flows in such terrain (Gal-Chen and Somerville 1975). Perhaps the most typical kind of transformation is one that converts the widely used cartesian form into a terrain-following one. One such transformation is as follows:

$$\sigma = H\left(\frac{z - z_g}{H - z_g}\right) \qquad (1.34)$$

where σ is the transformed vertical coordinate, H is the height of the top of the model domain, z is the height above mean sea level, z_g is the height of the ground above mean sea level and is a function of x and y. Transformations of this and similar kinds have been powerful tools in the analysis of meso-scale airflows over irregular terrain. They allow the dynamics outlined in this chapter to be applied to more complex and realistic flows than hitherto (eg Clark 1977).

1.13 Conclusion

This chapter has given only a brief outline of the basic dynamics of mesoscale airflows. After a consideration of the definition of the meso-scale, the dynamical equations have been investigated. It has been shown that the geostrophic approximation is valid only for airflows with horizontal length scales of the order of a thousand kilometres and more. The hydrostatic approximation is also a mainstay of large scale meteorology but has also been extensively used in meso-scale modelling. In the past two decades non-hydrostatic meso-scale models have become far more commonplace. The anelastic and Boussinesq approximations are widely used in analyses of meso-scale airflows. Many meso-scale circulations either contain parts of or are significantly influenced by the planetary boundary layer. This requires the behaviour of that layer to be carefully treated in the models which in turn requires understanding of the roles of elements such as surface fluxes, surface roughness, thermal stratification, radiation and moisture. This is a large field and the reader is referred to the works by Garratt (1992) and McBean (1979) for comprehensive coverage. Several aspects of the planetary boundary layer are also covered in other chapters in this volume.

References

Atkinson, B. W. 1981 *Meso-scale atmospheric circulations.* Academic Press 495pp

Blackadar, A.K. 1962 The vertical distribution of wind and turbulent exchange in the neutral atmosphere. *J. Geophys. Res.* **67**, 3095-3102.

Boussinesq, J. 1903 *Theorie analytique de la chaleur.* Gathier-Villars, Vol 2, 625pp

Businger, J.A., Wyngaard, J.C., Izumi,Y. and Bradley, E.F. 1971 Flux profile relationships in the atmospheric surface layer. *J. Atmos. Sci.* **28**, 181-189.

Clark, T.L. 1977 A small-scale dynamic model using a terrain-following coordinate transformation. *J. Comput. Phys.* **24**, 186-215.

Dutton, J.A. & Fichtl, G.H. 1969 Approximate equations of motion for gases and liquids. *J. Atmos. Sci.* **26**, 241-254.

Dyer, A.J. 1974 A review of flux-profile relationships. *Bound. Lay. Meteor.* **7**, 363-372.

Dyer, A.J. & Hicks, B.B. 1970 Flux-gradient relationships in the constant flux layer. *Quart. J. Roy. Met. Soc.* **96**, 715-721.

Estoque, M. A & Bhumralkar, C.M. 1969 Flow over a localized heat source. *Mon. Weath. Rev.* **97**, 850-859

Gal-Chen, T. & Somerville, R.C.J. 1975 On the use of a coordinate transformation for the solution of the Navier-Stokes equations. *J. Comput. Phys.* **17**, 209-228.

Garratt, J.R. 1992 *The atmospheric boundary layer.* Cambridge University Press 316 pp.

Gill, A. E. 1982 *Atmosphere-ocean dynamics.* Academic Press, 662 pp.

Gray, D.D. & Giorgini, A. 1976 The validity of the Boussinesq approximation for liquids and gases. *Int. J. Heat Mass Transfer* **19**, 545-551.

Hess, S.L. 1959 *Introduction to theoretical meteorology.* Holt, Rinehart and Winston 362pp

Ligda, M. G. H. 1951 Radar storm detection. In *Compendium of meteorology.* Ed T. F. Malone. American Meteorological Society, 1265-1282.

Lilly, D. K. 1983 Stratified turbulence and the meso-scale variability of the atmosphere. *J. Atmos. Sci.* **40**, 749-761.

Martin, C.L. & Pielke, R. A. 1983 The adequacy of the hydrostatic assumption in sea-breeze modelling over flat terrain. *J. Atmos. Sci.* **40**, 1472-1481.

Mc Bean, G.A. (Ed) 1979 The planetary boundary layer. *Tech. Note No. 165*, World Meteorological Organization 201 pp.

Mellor, G.L. & Yamada, T. 1974 A hierarchy of turbulence closure models for the planetary boundary layer. *J. Atmos. Sci.* **31**, 1791-1806.

Miller, M.J. & White, A.A. 1984 On the non-hydrostatic equations in pressure and sigma coordinates. *Quart. J. Roy. Meteor. Soc.* **110**, 515-534.

Monin, A.S. & Obukhov, A.M. 1954 Basic laws of turbulent mixing in the atmosphere near the ground. *Tr. Akad. Nauk, SSSR Geophiz. Inst.*, No. **24**(151), 1963-1987.

O'Brien, J.J. 1970 A note on the vertical structure of the eddy exchange cefficient in the planetary boundary layer. *J. Atmos. Sci.* **27**, 1213-1215.

Ogura, Y. & Phillips, N. 1962 Scale analysis of deep and shallow convection in the atmosphere. *J. Atmos. Sci.* **19**, 173-179.

Paulson, C.A. 1970 The mathematical representation of wind speed and temperature profiles in the unstable atmospheric surface layer. *J. Appl. Meteor.* **9**, 857-861.

Pielke, R. A. 1984 *Mesoscale meteorological modelling.* Academic Press 612pp

Smagorinsky, J. 1981 Epilogue: a perspective of dynamical meteorology. In *Dynamical meteorology. An introductory selection.* Ed. B. W. Atkinson, Methuen, 205-219.

Spiegal, E.A. & Veronis,G. 1960 On the Boussinesq approximation for a compressible fluid. *Astrophys. J.* **131**, 442-447.

Stull, R. B. 1988 *An introduction to boundary layer meteorology.* Kluwer Academic Publishers 666pp

Tepper, M. 1959 Mesometeorology - the link between macro-scale atmospheric motions and local weather. *Bull. of Amer. Meteor. Soc*, **40**, 56-72.

Vinnichenko, N. K. 1970 The kinetic energy spectrum in the free atmosphere - 1 second to 5 years. *Tellus* **22**, 158-166.

Yu, T-W. 1977 A comparative study on parameterization of vertical turbulent exchange processes. *Mon. Weath. Rev.* **105**, 57-66.

Address of the author

Prof B. W. Atkinson
Department of Geography
Queen Mary and Westfield College
University of London
Mile End Road
London E1 4NS
United Kingdom
e-mail address B.W.Atkinson@qmw.ac.uk

II Boundary Conditions and Treatment of Topography in Limited-Area Models

Dieter P. Eppel, Ulrich Callies

2.1. Introduction

Numerical simulations of atmospheric flows are based on the conservation principles of mass, momentum and energy. These principles - expressed in terms of conserved integrals - are conveniently cast into differential form, the hydrodynamical equations. Depending on the application, these equations are altered to a variety of different forms with different mathematical properties and different requirements on initial and/or boundary conditions in order to represent well-posed problems.

The notion of well-posedness has been introduced by Hadamard (1921) to characterize problems having solutions which depend smoothly on their initial and boundary values. For linear equations strong theorems on existence and uniqueness of solutions can be given.

For the hydrodynamical equations , in general existence of solutions can be proven only for limited time intervals, and uniqueness of the solution may be lost. One then has to select solutions which fullfill extra physical requirements, e.g. for the Euler Equations entropy conditions. It turns out that it is the well-posed problems which are amenable to stable numerical solutions, and for simple cases accuracy estimates can be given. As the mathematical proofs rarely lead to practical rules to treat the more complex applications, one is largely dependent on physical intuition and numerical experimentation.

Due to their origin the hydrodynamical equations describe a wealth of phenomena taking place on spacial and temporal scales covering many orders of magnitude. It is therefore useful and even necessary to investigate limiting cases of the general equations. In the context here, the low Mach Number regime is of special interest. Though the atmosphere behaves in many respects as an incompressible fluid, buoyancy is one of the mayor driving forces, and it can hardly ever be neglected. So the discussion focusses on low Mach Number flow of stratified fluids.

In the remainder of this section some simplified versions of the equations are presented which are used in high-resolution atmospheric flow modelling, and for which ultimately initial- and boundary conditions are sought. In Section 2 a short listing of nomenclature and elementary properties of the basic types of equations is given to serve as guideline. In section 3 numerical techniques for treating boundaries are discussed. Naturally, the greatest difficulties are posed by open boundaries. Their treatment makes up for most of the presentation. In the final section 4, problems related to the presence of topographical variations are described. Citations from

A. Gyr and F-S. Rys (eds.), Diffusion and Transport of Pollutants in Atmospheric Mesoscale Flow Fields, 23–56.

the vast literature are by far not exhaustive. However, they should serve as starting point for closer studies.

Basic Equations

The discussion centers around the set of partial differential equations used to describe the airflow in a limited domain of the lower troposphere, the surface of the earth being the only solid boundary. The time evolution of momentum and mass are governed by the *Navier-Stokes Equations* and by the *continuity equation*:

$$\begin{aligned} \rho\frac{d}{dt}\mathbf{u} + \nabla p - \mu_m \Delta\mathbf{u} &= \rho\mathbf{g} - 2\rho\boldsymbol{\Omega}\times\mathbf{u} \\ \frac{d}{dt}\rho + \rho\nabla\cdot\mathbf{u} &= 0 \end{aligned} \tag{2.1.1}$$

ρ is the density, $\mathbf{u}$ denotes the velocity field, p is the pressure, $\mathbf{g}$ the gravity vector, and $\boldsymbol{\Omega}$ the Coriolis Vector. The stress divergence which, upon closer inspection, contains the effects of subgrid-scale interactions not explicitly treated, has been approximated by a simple Laplacian in order to have second order derivatives present.

Thermodynamics is coupled into the system by an equation of state (ideal gas) and the *First Law* of Thermodynamics ($\alpha = 1/\rho$):

$$\begin{aligned} p\alpha &= RT \\ Tds &= de + pd\alpha = 0 \end{aligned} \tag{2.1.2}$$

s and e are specific entropy and specific inner energy, respectively. $T\,ds$ is the specific heat change.

Choosing T, α as independent variables (i.e. $s = s(T,\alpha)$, $e = e(T,\alpha)$, equation (2.1.2) can be rewritten in the form :

$$ds = \frac{c_V}{T}dT + \frac{R}{\alpha}d\alpha = c_V d(\ln T) + Rd(\ln\alpha) \tag{2.1.3}$$

with

$$c_V = T\frac{\partial s}{\partial T}\bigg|_\alpha, \quad R = c_p - c_V\,. \tag{2.1.4}$$

shows that ds is an *exact differential*. Therefore, s can be chosen as a dynamical variable. It is customary in meteorology to use instead the *potential temperature*, θ, which is defined through the integrated form of eq. (2.1.3):

$$s - s_0 = c_p \ln\left[\left(\frac{T}{T_0}\right)\left(\frac{p_0}{p}\right)^{\frac{R}{c_p}}\right] =_{def} c_p \ln\left(\frac{\theta}{\theta_0}\right)\,. \tag{2.1.5}$$

Choosing $\theta_0 = T_0$ one has

$$\theta = T\left(\frac{p_0}{p}\right)^{\frac{R}{c_p}}\,. \tag{2.1.6}$$

From equation (2.1.5) also follows an expression of the sound speed, c_s. For constant entropy, $s = s_0$, together with the equation of state (2.1.2) one has

$$c_s^2 = \left(\frac{\partial p}{\partial \rho}\right)_s = \gamma\frac{p}{\rho} = \gamma RT\,, \quad \gamma = \frac{c_p}{c_V}\,. \tag{2.1.7}$$

According to equation (2.1.5) heat supply (either reversible or irreversible) is the only way to change the potential temperature of a given fluid parcel:

$$c_p\frac{T}{\theta}\frac{d\theta}{dt} = T\frac{ds}{dt}\,. \tag{2.1.8}$$

The only diabatic process in the example here will be heat conduction giving rise to a diffusive entropy flux density $\mathbf{J}_s$. This flux, $\mathbf{J}_s$, redistributes the actual entropy and causes, at the same time, local production of entropy. This double role is reflected in the entropy balance equation where $\mathbf{J}_s$ occurs twice in different terms (see e.g. Gyarmati, 1970):

$$\frac{\partial}{\partial t}(\rho s) + \nabla\cdot(\mathbf{u}\rho s + \mathbf{J}_s) = \mathbf{J}_q\cdot\nabla\left(\frac{1}{T}\right)\,, \quad \mathbf{J}_q = T\mathbf{J}_s\,. \tag{2.1.9}$$

A positive sign of the source term on the r.h.s. of equation (2.1.9) must be postulated according to the *Second Law* of Thermodynamics.

Here, it is assumed that a simple Fourier law for heat conduction is valid

$$\mathbf{J}_q = T\mathbf{J}_s = \mu_T\nabla T \tag{2.1.10}$$

with constant *thermal conductivity*, μ_T. Then, using the continuity equation from (2.1.1) the equation (2.1.9) can be cast into the form:

$$\rho T\frac{ds}{dt} = \mu_T\Delta T \tag{2.1.11}$$

where the thermal conductivity, μ_T, is related to the thermal diffusivity, ν_T, by $\mu_T = \rho c_p \nu_T$. The basic equations are therefore:

$$\begin{aligned}
\rho\frac{d\mathbf{u}}{dt} + \nabla p - \mu_m\Delta\mathbf{u} &= \rho\mathbf{g} - 2\rho\boldsymbol{\Omega}\times\mathbf{u} \\
\frac{d\rho}{dt} + \rho\nabla\cdot\mathbf{u} &= 0 \\
\rho\frac{d\theta}{dt} &= \frac{\mu_T}{c_p}\frac{\theta}{T}\Delta T \\
p\alpha = RT\,, \qquad \theta &= T\left(\frac{p_0}{p}\right)^{\frac{R}{c_p}}\,.
\end{aligned} \tag{2.1.12}$$

Equations (2.1.12) are given in *quasi-linear form* which is preferred in theoretical investigations. For discrete approximations the analytically equivalent *flux-form* is often used where the *advective terms* appear as divergences,

$$\rho\frac{d}{dt}\mathbf{u} \quad\leftrightarrow\quad \frac{\partial}{\partial t}(\rho\mathbf{u}) + \nabla\cdot(\mathbf{u}\otimes\rho\mathbf{u})\,, \tag{2.1.13}$$

because this form guarantees exact conservation of the quantity during discrete advection. The flux-form of eqs. (2.1.12) is:

$$\begin{aligned} \frac{\partial}{\partial t}(\rho\mathbf{u}) + \nabla\cdot(\mathbf{u}\otimes\rho\mathbf{u}) + \nabla p - \mu_m\Delta\mathbf{u} &= \rho\mathbf{g} - 2\rho\boldsymbol{\Omega}\times\mathbf{u} \\ \frac{\partial}{\partial t}\rho + \nabla\cdot(\mathbf{u}\rho) &= 0 \\ \frac{\partial}{\partial t}(\rho\theta) + \nabla\cdot(\mathbf{u}\rho\theta) &= \frac{\mu_T}{c_p}\frac{\theta}{T}\Delta T \\ p\alpha = RT\,, \qquad \theta &= T\left(\frac{p_0}{p}\right)^{\frac{R}{c_p}} \end{aligned} \tag{2.1.14}$$

(the symbol "$\otimes$" denotes the tensor product of two vectors).

A reasonable choice among others for the integration constants, p_0, T_0 and ρ_0, is their sea-level values of a standard atmosphere: $p_0 = 101.325$ hPa, $T_0 = 288.15\ ^\circ$ K, $\rho_0 = 1.225$ kg/m^3. The other constants are $R = 287$ J/(K kg), $c_p = 1004$ J/(K kg), $c_V = 717$ J/(K kg). These values can also be taken for scaling the equations.

Scaling

Scaling gives insight into the relative sizes of the different terms in the equations thus indicating which phenomema are to be expected. The choice of algorithms and the accuracy of the numerical solution can depend on the chosen scaling.

The velocity field, gradient and Laplacian are separated into horizontal and vertical contributions (the superscript T denotes transposition),

$$\mathbf{u} = (\mathbf{u}_H, w)^T = (u, v, w)^T, \qquad \nabla = (\nabla_H, \frac{\partial}{\partial z})^T\,, \qquad \Delta = \Delta_H + \frac{\partial^2}{\partial z^2}\,, \tag{2.1.15}$$

and the substantial time derivatives are written in Eulerian form. For simplicity, only the vertical component of the Coriolis force is taken into account: $\boldsymbol{\Omega} = (0, 0, \Omega_3)$. The gravity vector is $\mathbf{g} = (0, 0, -g)$. The following scaling is introduced:

$$\begin{array}{lllll} x = l_0\hat{x}\,, & z = h_0\hat{z}\,, & t = t_0\hat{t}\,, & u = u_0\hat{u}\,, & w = w_0\hat{w}\,, \\ T = T_0\hat{T}\,, & \theta = \theta_0\hat{\theta}\,, & \theta_0 = T_0\,, & \rho = \rho_0\hat{\rho}\,, & p = p_0\hat{p} \end{array} \tag{2.1.16}$$

such that the quantities with hats are $O(1)$. As the advective terms should be invariant to scaling because they represent a geometrical property of the flow one requires

$$\frac{u_0 t_0}{l_0} = \frac{w_0 t_0}{h_0} = 1 \quad \rightarrow \quad u_0 = \frac{l_0}{t_0}\,, \quad w_0 = \frac{h_0}{t_0}\,. \tag{2.1.17}$$

Therefore, vertical and horizontal velocity scales are related through their common time scale by which the aspect ratio α is fixed:

$$\alpha =_{\text{def}} \frac{h_0}{l_0} = \frac{w_0}{u_0}\,. \tag{2.1.18}$$

This is a purely kinematical scaling. If we were to scale, for example, the vertical velocity w by Deardorff/s (1972) convective velocity scale, $w_* = (Q_0 z_i g/\theta_0)^{1/3}$ (Q_0 is the heat flux, z_i the height of the convective layer), the relation (2.1.18) would fix the vertical scale h_0.

Taking the choice (2.1.17) for the equations, (2.1.12) yields ($\nu_m = \mu_m/\rho_0, \nu_T = \mu_T/(\rho_0 c_p)$):

$$\begin{aligned}
\hat{\rho}\left[\frac{\partial}{\partial \hat{t}} + \hat{\mathbf{u}}\cdot\hat{\nabla} - \frac{\nu_m}{h_0 w_0}\left(\alpha^2\hat{\Delta}_H + \frac{\partial^2}{\partial \hat{z}^2}\right)\right]\hat{\mathbf{u}}_H &= -\frac{p_0}{\rho_0 u_0^2}\hat{\nabla}_H\hat{p} - 2\hat{\rho}t_0\left\{\begin{array}{c}-\Omega_3\hat{v}\\ \Omega_3\hat{u}\end{array}\right\}, \\
\hat{\rho}\left[\frac{\partial}{\partial \hat{t}} + \hat{\mathbf{u}}\cdot\hat{\nabla} - \frac{\nu_m}{h_0 w_0}\left(\alpha^2\hat{\Delta}_H + \frac{\partial^2}{\partial \hat{z}^2}\right)\right]\hat{w} &= -\frac{p_0}{\rho_0 w_0^2}\frac{\partial}{\partial \hat{z}}\hat{p} - \frac{t_0}{w_0}\hat{\rho}g, \\
\left[\frac{\partial}{\partial \hat{t}} + \hat{\mathbf{u}}\cdot\hat{\nabla}\right]\hat{\rho} &= -\hat{\rho}[\hat{\nabla}\cdot\hat{\mathbf{u}}], \\
\hat{\rho}\left[\frac{\partial}{\partial \hat{t}} + \hat{\mathbf{u}}\cdot\hat{\nabla}\right]\hat{\theta} - \left(\frac{\nu_T}{h_0 w_0}\right)\frac{\hat{\theta}}{\hat{T}}\left(\alpha^2\hat{\Delta}_H + \frac{\partial^2}{\partial \hat{z}^2}\right)\hat{T} &= 0, \\
\hat{p} &= (\hat{\rho}\hat{\theta})^\gamma .
\end{aligned} \tag{2.1.19}$$

Introducing the quantities

$$\begin{aligned}
Eu^2 &= \frac{\rho_0 u_0^2}{p_0} = \frac{\gamma u_0^2}{(\partial p_0/\partial \rho_0)_s} = \gamma\left(\frac{u_0}{c_s}\right)^2 && \text{Euler Number} \\
Re &= \frac{h_0 w_0}{\nu_m} && \text{Reynolds Number} \\
Pr &= \frac{\nu_m}{\nu_T} && \text{Prandtl Number} \\
Ro &= \frac{2\Omega_3 l_0}{u_0} && \text{Rossby Number} \\
Fr &= \frac{w_0}{g t_0} = \frac{w_0^2}{g h_0} && \text{Froude Number}
\end{aligned} \tag{2.1.20}$$

the final scaled equations are then:

$$\begin{aligned}
\hat{\rho}\left[\frac{\partial}{\partial \hat{t}} + \hat{\mathbf{u}}\cdot\hat{\nabla} - \frac{1}{Re}\left(\alpha^2\hat{\Delta}_H + \frac{\partial^2}{\partial \hat{z}^2}\right)\right]\hat{\mathbf{u}}_H &= -\frac{1}{Eu^2}\hat{\nabla}_H\hat{p} - \frac{1}{Ro}\mathbf{f}\times\hat{\mathbf{u}}_H, \\
\hat{\rho}\left[\frac{\partial}{\partial \hat{t}} + \hat{\mathbf{u}}\cdot\hat{\nabla} - \frac{1}{Re}\left(\alpha^2\hat{\Delta}_H + \frac{\partial^2}{\partial \hat{z}^2}\right)\right]\hat{w} &= -\frac{1}{\alpha^2 Eu^2}\frac{\partial}{\partial \hat{z}}\hat{p} - \frac{1}{Fr}\hat{\rho}, \\
\left[\frac{\partial}{\partial \hat{t}} + \hat{\mathbf{u}}\cdot\hat{\nabla}\right]\hat{\rho} &= -\hat{\rho}[\hat{\nabla}\cdot\hat{\mathbf{u}}], \\
\hat{\rho}\left[\frac{\partial}{\partial \hat{t}} + \hat{\mathbf{u}}\cdot\hat{\nabla}\right]\hat{\theta} - \frac{1}{(RePr)}\frac{\hat{\theta}}{\hat{T}}\left(\alpha^2\hat{\Delta}_H + \frac{\partial^2}{\partial \hat{z}^2}\right)\hat{T} &= 0, \\
\hat{p} &= (\hat{\rho}\hat{\theta})^\gamma .
\end{aligned} \tag{2.1.21}$$

In the form (2.1.21) the physical forces are identified by their nondimensional numbers. The highest derivatives do not occur in all equations. The system is therefore of *mixed type* in the classification described below. Only for high-enough Reynolds Numbers do all equations belong to the same type of first-order hyperbolic equations. Numerical solutions should always be sought from the properly scaled equations.

Background States

It is customary not to use the equations directly but to calculate the deviations from a known background state, and there are arguments to carefully select this state (see e.g. Cullen, 1990). The most common background states among others are an isentropic atmosphere and an isothermal atmosphere.

Isentropic Atmosphere. The *isentropic background* state is defined by setting $s = \overline{s} = s_0$. Denoting the isentropic values of pressure, density and temperature by overbars, $\overline{p}$, $\overline{\rho}$, $\overline{T}$, and assuming mechanical balance of the atmosphere,

$$\frac{\partial}{\partial z}\overline{p}(z) = -\overline{\rho}(z)g\ , \tag{2.1.22}$$

one obtains:

$$\overline{p}(z) = p_0 h(z)^{\frac{c_p}{R}}, \quad \overline{\rho}(z) = \rho_0 h(z)^{\frac{c_V}{R}}, \quad \overline{T}(z) = T_0 h(z), \quad h(z) = 1 - \frac{z}{H_s}\ . \tag{2.1.23}$$

The linear decrease of T with height implies $\theta(z) = \theta_0 =$ const. The isentropic scale height, H_s, is given by

$$H_s = \frac{c_p T_0}{g}\ . \tag{2.1.24}$$

For the standard atmosphere $H_s = 288.15 \cdot 1004/9.81 = 29491\ m$. It should be noted that an isentropic atmosphere has a finite height.

Isothermal Atmosphere. In this case $\overline{T} = T_0$. Together with mechanical balance (eq. (2.1.20)) one obtains for the *isothermal background* state

$$\overline{p}(z) = p_0 e^{-z/H_\rho}\ , \qquad \overline{\rho}(z) = \rho_0 e^{-z/H_\rho}\ , \qquad \overline{\theta}(z) = \theta_0 e^{z/H_\theta} \tag{2.1.25}$$

with

$$H_\rho = \frac{RT_0}{g}\ , \qquad H_\theta = \frac{c_p T_0}{g}\ . \tag{2.1.26}$$

It is customary to choose $T_0 = 300°$K. For the scale heights follows $z_\rho = 8777$ m, and $z_\theta = 30703$ m. Keeping the reference pressure $p_0 = 101.325$ hPa results in a reference density of $\rho_0 = 1.176$ kg/m^3.

Approximations

The dynamical equations are rarely used in their bare form. Rather one derives for the scales considered equations in which contributions considered small are neglected.

Bousinesq Approximation. The *Bousinesq Approximation*, resulting from the notion that density deviations in the atmosphere from their background state are to a high degree balanced by temperature deviations, serves to simplify the buoyancy term. Separating $p = \overline{p} + p'$, etc. the Boussinesq approximation is obtained by using

$$\frac{p'}{\overline{p}} \simeq \gamma \left[\frac{\rho'}{\overline{\rho}} + \frac{\theta'}{\overline{\theta}} \right] \simeq 0 \tag{2.1.27}$$

to relate density variations in the buoyancy term of the momentum equation to temperature only, i.e. density deviations are taken into account only in connection with gravity. The set of equations in Boussinesq approximation then is:

$$\begin{aligned} \overline{\rho}\frac{d}{dt}\mathbf{u} + \nabla p' - \mu_m \Delta \mathbf{u} &= -2\overline{\rho}\mathbf{\Omega} \times \mathbf{u} - \frac{\overline{\rho}}{\overline{\theta}}\mathbf{g}\theta' \\ \frac{d}{dt}\rho + \rho \nabla \cdot \mathbf{u} &= 0 \\ \overline{\rho}\frac{d}{dt}\theta &= \frac{\mu_T}{c_p}\frac{\theta}{T}\Delta T \end{aligned} \tag{2.1.28}$$

Incompressibility Limits. The limit to so-called *anelastic incompressibility* is obtained by the following limiting procedure in the continuity equation (2.1.28) ($\rho = \overline{\rho} + \epsilon\rho'$)

$$\begin{aligned} \frac{\partial}{\partial t}(\overline{\rho} + \epsilon\rho') + (\overline{\rho} + \epsilon\rho')\nabla \cdot \mathbf{u} + \mathbf{u} \cdot \nabla(\overline{\rho} + \epsilon\rho') &= 0 \\ \epsilon \left(\frac{\partial}{\partial t}\rho' + \mathbf{u} \cdot \nabla\rho' + \rho'\nabla \cdot \mathbf{u} \right) + \mathbf{u} \cdot \nabla\overline{\rho} + \overline{\rho}\nabla \cdot \mathbf{u} &= 0 \end{aligned}$$

For small enough ϵ one gets

$$\mathbf{u} \cdot \nabla\overline{\rho} + \overline{\rho}\nabla \cdot \mathbf{u} = 0 \tag{2.1.29}$$

Similarly, the transition $\epsilon \longrightarrow 0$ in the momentum and temperature equations is obtained. One finally has the anelastic approximation,

$$\begin{aligned} \overline{\rho}\frac{d}{dt}\mathbf{u} + \nabla p' - \mu_m \Delta \mathbf{u} &= -2\overline{\rho}\mathbf{\Omega} \times \mathbf{u} - \frac{\overline{\rho}}{\overline{\theta}}\mathbf{g}\theta' \\ \nabla \cdot (\overline{\rho}\mathbf{u}) &= 0 \\ \overline{\rho}\frac{d}{dt}\theta &= \frac{\mu_T}{c_p}\frac{\theta}{T}\Delta T \\ \theta &= T\left(\frac{p_0}{p}\right)^{\frac{R}{c_p}} \end{aligned} \tag{2.1.30}$$

Equations (2.1.30) have been derived within a rigorous scale analysis (Ogura and Phillips, 1962). It can be shown that this set of equations does not support sound waves but the gravity waves are still present with their dispersion relations being only marginally different from those obtained from the fully compressible equations (Wippermann, 1981). Moreover, in eqs. (2.1.30) energy is conserved to the same order as the original approximation (Dutton and Fichtl, 1969). If instead, an isothermal background state is chosen the incompressibility condition in (2.1.30) should be replaced by (Durran, 1989):

$$\nabla \cdot (\overline{\rho}\overline{\theta}\mathbf{u}) = \frac{H}{c_p}\left(\frac{p}{p_0}\right)^{-R/c_p} = \frac{H}{c_p}\frac{\theta}{T}$$

where H is the volume heating rate.

An even simpler form is obtained when expanding around a spacially constant density. In this case one uses

$$\frac{p'}{p_0} = \frac{\rho'}{\rho_0} + \frac{T'}{T_0} \simeq 0\,. \tag{2.1.31}$$

The set of equations then simply reduces to

$$\begin{aligned}
\rho_0\frac{d}{dt}\mathbf{u} + \nabla p' - \mu_m \Delta \mathbf{u} &= -2\rho_0 \mathbf{\Omega} \times \mathbf{u} - \frac{\rho_0}{T_0}\mathbf{g}T' \\
\nabla \cdot \mathbf{u} &= 0 \\
\rho_0 \frac{d}{dt}T &= \frac{\mu_T}{c_V}\Delta T\,.
\end{aligned} \tag{2.1.32}$$

A rigorous limiting procedure for nonstratified flow has been given by Klainerman and Majda (1981, 1982).

Due to the approximations (2.1.27) and (2.1.31) there is no equation-of-state relating pressure to density and temperature to close the equation system. Pressure itself has to be determined implicitly such that the incompressibility condition is fulfilled.

2.2 Types of Equations

The equations summarized in the last section require different boundary conditions depending on the type they belong to. As only first and second derivatives naturally occur in the equations for airflow a short listing of basic facts on first-order hyperbolic and second-order parabolic and elliptic equations is given. An extensive discussion relating to the *Navier-Stokes Equations* can be found in Kreiss and Lorenz (1989) from where much of the material below is drawn.

It is essentially two problem areas one should be aware of. First, there is the mathematical notion of *well-posedness.* There exist a variety of definitions of well-posedness in the literature. Here, we adhere to the form which has direct implications on the numerical treatment of the hydrodynamical equations, and which is related to Hadamard's (1921) definition. If a problem can be formulated as a well-posed problem then there is the second problem of technically or numerically implementing the mathematical boundary conditions. Moreover, since the theory for the NSE (and therefore also for the hydrodynamical equations) is not completely settled it needs numerical experimentation to arrive at satisfactory numerical models.

The *Navier-Stokes Equations* form a quasilinear differential system, and much of our understanding is gained through the study of linearized equations. These, in general, will have variable coefficients. By "freezing" the coefficients one obtains systems with constant coefficients which are easier to investigate. Therefore, investigation of the well-posedness of the Navier-Stokes Equations proceeds along a sequence of increasingly more complicated equations. Roughly speaking the approach goes as follows: first, it is proven that the Cauchy problem (i.e. the pure initial-value problem, boundary conditions are avoided by prescribing spacial periodicity) for systems with constant coefficients is well-posed. This part is 'simple' insofar as the problem can be reduced to the investigation of Fourier modes. Then one can prove that if all systems with constant coefficients resulting from linearizations of a linear system with non-constant coefficients are well-posed then the non-constant coefficient case is well-posed. This latter statement is not true for arbitrary equations but the linearized Navier-Stokes Equations fall into this category. The next step then is to prove that if all problems with non-constant coefficients resulting from the linearization of a non-linear problem are well-posed then the non-linear problem is well-posed. However, one has lost the validity for all times: there is only a limited time interval within which well-posedness can be proven.

These results can be taken over more or less to initial/boundary value problems of the Navier-Stokes type. In the following a few definitions and theorems are cited to show the kind of knowledge which results from the mathematical investigations.

To be more precise some nomenclature is needed and, above all, a norm has to be specified. Though mostly valid for arbitrary dimensions the statements are written for two- or three-dimensional space. A domain in $\mathbf{R}^3$ (or $\mathbf{R}^2$) is denoted by $\boldsymbol{\Omega}$ with its (sufficiently smooth) boundary $\boldsymbol{\Gamma}$. The spacial vector is denoted by $\mathbf{x} \in \mathbf{R}^3$. The variables of the problem are arranged as a vector:

$$\left\{ \begin{matrix} u_1(\mathbf{x},t) \\ \dots \\ u_n(\mathbf{x},t) \end{matrix} \right\} = \mathbf{u}(\mathbf{x},t) \tag{2.2.1}$$

with the *scalar product* defined as

$$\begin{aligned} (\mathbf{u},\mathbf{v}) &= \sum_{j=1}^{n} \int_{\boldsymbol{\Omega}} u_j(\mathbf{x},t) v_j(\mathbf{x},t)\, d\mathbf{x}\,, \qquad \|\mathbf{u}\|^2 = (\mathbf{u},\mathbf{u}) \\ (\mathbf{u},\mathbf{v})_{\boldsymbol{\Gamma}} &= \sum_{j=1}^{n} \int_{\boldsymbol{\Gamma}} u_j(\mathbf{x},t) v_j(\mathbf{x},t)\, d\mathbf{x}\,, \qquad \|\mathbf{u}\|_{\boldsymbol{\Gamma}}^2 = (\mathbf{u},\mathbf{u})_{\boldsymbol{\Gamma}} \end{aligned} \tag{2.2.2}$$

There are other norms which may serve as definition for well-posedness, e.g. energy norms. The l^2-norm is commonly used for estimates of finite-difference approximations to the equations. It is important to note that, in the definition of the norm, no derivative terms occur. Moreover, the equations have to be in nondimensional form to apply the mathematical definitions and theorem.

Well-Posedness

Suppose a quasilinear system of PDE's is given (quasilinear means that the equations are linear in the derivatives) in a domain $\boldsymbol{\Omega}$ with boundary $\boldsymbol{\Gamma}$ of $\mathbf{R}^3$ and with initial and boundary conditions:

$$\begin{aligned} \mathbf{u}_t = \mathbf{P}(\mathbf{x},t,\mathbf{u},\partial/\partial x_j)\mathbf{u} + \mathbf{F}(\mathbf{x},t)\,, &\quad \mathbf{x}\in\boldsymbol{\Omega}\,, \quad 0\le t\le T\,, \\ \mathbf{u}(\mathbf{x},0) = \mathbf{f}(\mathbf{x})\,, &\quad \mathbf{x}\in\boldsymbol{\Omega}\,, \\ \mathbf{L}\mathbf{u} = \mathbf{g}(\mathbf{x},t)\,, &\quad \mathbf{x}\in\boldsymbol{\Gamma}\,. \end{aligned} \tag{2.2.3}$$

The operator $\mathbf{P}$ is a *quasilinear differential operator* of order m and has the general form:

$$\mathbf{P}(\mathbf{x},t,\mathbf{u},\partial/\partial x_j) = \sum_{\nu\le 2} \mathbf{A}_\nu(\mathbf{x},t,\mathbf{u}) \frac{\partial^\nu}{\partial x_1^{\nu_1}\ldots\partial x_3^{\nu_3}}\,. \tag{2.2.4}$$

The index ν is given as $\nu = \nu_1 + \ldots + \nu_3$, $(0 \le \nu_i)$. The coefficients $\mathbf{A}_\nu$ are matrices, $\mathbf{A}_\nu \in \mathbf{C}^{n,n}$ and smooth functions of their arguments. The same shall hold for the initial and boundary conditions, $\mathbf{f}$, $\mathbf{g}$, and for the forcing, $\mathbf{F}$. The boundary operator $\mathbf{L}$ is assumed to be linear. The initial and boundary conditions shall not contradict each other.

From a purely mathematical point of view there are two fundamental questions:

- when has eq. (2.2.3) a unique solution, at least for a small time interval?
- Assume eq. (2.2.3) has a unique solution. What is the influence of perturbations of $\mathbf{F}$, f and g on the solution.

The first question is answered by the Cauchy-Kowalewskaya Theorem stating that if all the functions occurring in eq. (2.2.3) are analytic functions of their arguments then a unique solution in a neighbourhood of the initial condition exists. However, nothing is said about the quality of the solution.

The second question about the behaviour of the solution can be cast into a definition for well-posedness: perturb the data, $\mathbf{f}$, $\mathbf{g}$ and $\mathbf{F}$ in eq. (2.2.3) by $\delta\mathbf{f}$, $\delta\mathbf{g}$ and $\delta\mathbf{F}$, and consider the *perturbed system*

$$\begin{aligned} \mathbf{v}_t = \mathbf{P}(\mathbf{x},t,\mathbf{v},\partial/\partial x_j)\mathbf{v} + \mathbf{F} + \delta\mathbf{F}\,, &\quad \mathbf{x}\in\boldsymbol{\Omega}\,, \quad 0\le t\le T\,, \\ \mathbf{v}(\mathbf{x},0) = \mathbf{f} + \delta\mathbf{f}\,, &\quad \mathbf{L}\mathbf{v} = \mathbf{g} + \delta\mathbf{g}\,. \end{aligned} \tag{2.2.5}$$

The system is considered well-posed if, for small perturbations, the *perturbed solution* is close to the original solution, i.e with :

$$\mathbf{v} = \mathbf{u} + \delta\mathbf{u} \tag{2.2.6}$$

$\delta\mathbf{u}$ shall satisfy an estimate of the form:

$$\|\delta\mathbf{u}\| + \|\delta\mathbf{u}\|_\Gamma \le K(\|\delta\mathbf{F}\| + \|\delta\mathbf{f}\| + \|\delta\mathbf{g}\|_\Gamma)\,. \tag{2.2.7}$$

If (2.2.7) is fulfilled one says that the perturbation $\delta\mathbf{u}$ can be "estimated by the data" $\delta\mathbf{F}$, $\delta\mathbf{f}$ and $\delta\mathbf{g}$. Or, small perturbations create only small changes in the solution.

Classification of Equations

There is an intuitive physical classification of the different types of equations which is frequently used when devising boundary conditions. When calculating dispersion relations from the linearized equations (for which $\mathbf{u}(\mathbf{x},t) = \mathbf{u}_0 \exp(i\mathbf{k}\cdot\mathbf{x} - i\omega t)$ is an eigensolution) one obtains frequency as a function of wave vector, $\omega = \omega(\mathbf{k})$. If the phase speed, $\mathbf{v}_p = (\omega/k_1, \omega/k_2, \omega/k_3)^T$, is real the system is said to be hyperbolic; if the phase speed is complex with negative imaginary part then the system is parabolic, and if the phase speed is purely imaginary then the system is elliptic.

A more precise definition can be given with the help of the so-called symbol of the linearized differential operator $\mathbf{P}$ from equation (2.2.4). Introducing the abbreviation $D_j = \partial/\partial x_j$ the linearized constant-coefficient operator is:

$$\mathbf{P}(\frac{\partial}{\partial x_j}) = \sum_{\nu\leq 2} \mathbf{A}_\nu \frac{\partial^{|\nu|}}{\partial x_1^{\nu_1}\dots\partial x_3^{\nu_3}} = \sum_{\nu\leq 2} \mathbf{A}_\nu D_1^{\nu_1}\dots D_3^{\nu_3} \tag{2.2.8}$$

which, when applied to a Fourier component of the solution, transforms into

$$\begin{aligned}\mathbf{P}(ik_j) &= \sum_{\nu\leq 2} A_\nu (ik_1)^{\nu_1}\dots(ik_3)^{\nu_3} \\ &= \mathbf{P}_0(ik_j) + \mathbf{P}_1(ik_j) + \mathbf{P}_2(ik_j)\,.\end{aligned} \tag{2.2.9}$$

$\mathbf{P}(ik_j)$ is called the *symbol* of $\mathbf{P}(\partial/\partial x_j)$, and that part containing the highest derivatives, $\mathbf{P}_2$, is called its *principal part.*

The eigenvalues and eigensolutions of the symbol are used to classify the different types of equations.

Parabolic Equations. An equation system of second order

$$\frac{\partial}{\partial t}\mathbf{u} = \sum_{j'=0}^{2} \mathbf{P}_{j'}(\frac{\partial}{\partial x_j})\mathbf{u} \tag{2.2.10}$$

is called *parabolic* if the symbol of the principal part, $\mathbf{P}_2(ik_j)$, has only negative eigenvalues for all $\mathbf{k}$, i.e there is a $\delta > 0$:

$$\lambda_j(\mathbf{k}) \leq -\delta\,|\mathbf{k}|^2\,. \tag{2.2.11}$$

As the solution evolves in time according to

$$\mathbf{u}(t) \sim \int d\mathbf{k} e^{t\cdot\mathbf{P}_2(ik_j)} \left[e^{t\cdot(\mathbf{P}_1(ik_j)+\mathbf{P}_0(ik_j))}\right] \mathbf{u}_\mathbf{k}(0) \tag{2.2.12}$$

it will always decay irrespective of the lower-order terms. As equation (2.2.10) is linear, estimates of the solution in terms of the symbol using (2.2.12) are possible.

Linear parabolic problems with constant coefficients are well-posed under very general conditions. As already mentioned this property can be generalized to variable-coefficient nonlinear parabolic problems. The following theorem for linear parabolic constant-coefficient problems essentially covers the the parabolic approximations used in airflow modelling.

Given the problem

$$\begin{aligned} \mathbf{u}_t &= \sum_{i,j=1}^{3} \mathbf{A}_{ij} D_i D_j \mathbf{u} + \sum_{i=1}^{3} \mathbf{B}_i D_i \mathbf{u} + \mathbf{C}\mathbf{u} + \mathbf{F}\,, \\ \mathbf{u}(\mathbf{u},0) &= \mathbf{f}(\mathbf{x})\,, \qquad \mathbf{x} \in \boldsymbol{\Omega}\,, \\ a\mathbf{L}\mathbf{u}(\mathbf{x},t) &+ b\mathbf{D}_n \mathbf{u}(\mathbf{x},t) = \mathbf{g}(\mathbf{x},t)\,, \quad \mathbf{x} \in \boldsymbol{\Gamma}\,. \end{aligned} \tag{2.2.13}$$

Here, $\mathbf{u}$ is the vector of n dependent variables, e.g. $\mathbf{u} = (u, v, w, \theta, \ldots)^T$. The $\mathbf{A}_{ij}$, $\mathbf{B}_i$, $\mathbf{C}$ are $\mathbf{C}^{n,n}$ matrices, $\mathbf{D}_n$ denotes the normal derivative on the boundary, and $\mathbf{L}$ is a linear operator. Therefore, *Dirichlet boundary conditions* (values of the variables are prescribed on the boundary) and *Neumann boundary conditions* (values and/or derivatives are prescribed) are contained in the formulation. Since estimates on the solution are obtained by partial integration which generate boundary terms the forcing terms have to fulfill certain compatibility conditions. A sufficient condition is that all the forcing terms vanish at $t = 0$ in a small neighbourhood of the boundary. If, in addition, (2.2.13) is *strongly parabolic*, i.e.

$$\sum_{i,j=1}^{3} \langle y_i \mathbf{A}_{ij} y_j \rangle \geq \delta \sum_{i=1}^{3} |y_i|^2 \tag{2.2.14}$$

then eq. (2.2.13) is a well-posed problem. This means there exists a constant $K_T > 0$ independent of the 'data' that for a certain time interval, $0 \leq t \leq T$, the following estimate is valid:

$$\|\mathbf{u}(t)\|^2 + \int_0^t \|\mathbf{u}(\tau)\|_{\mathbf{\Gamma}}^2 \, d\tau \leq K_T \left\{ \|\mathbf{f}\|^2 + \int_0^t \left[\|\mathbf{F}(\tau)\|^2 + \|\mathbf{g}(\tau)\|_{\mathbf{\Gamma}}^2 \right] d\tau \right\} . \tag{2.2.15}$$

Irrespective of the generation of instabilities by *lower-order terms* the second-order operator asymptotically dominates and guarantees well-posedness.

Elliptic Equations. The equation for pressure in incompressible flow serves as a prototype of a scalar *elliptic equation*:

$$\begin{aligned} a_{ij} D_i D_j p(\mathbf{x}) + F(\mathbf{x}) = 0\,, &\quad \mathbf{x} \in \boldsymbol{\Omega}\,, \\ a p(\mathbf{x}) + b d_n p(\mathbf{x}) = g(\mathbf{x})\,, &\quad \mathbf{x} \in \boldsymbol{\Gamma}\,. \end{aligned} \tag{2.2.16}$$

If the matrix a_{ij} is definite (i.e. there exist as many real eigenvalues as there are space dimensions, and the eigenvalues have all the same sign) then the problem is well-posed. If the Dirichlet part of the boundary condition in eq. (2.2.16) is zero (a=0) then the solution is determined up to constant.

Hyperbolic Equations. As *hyperbolic systems* support free wave propagation often with more than one phase speed the formulation of well-posed boundary conditions is complicated by the need to identify and discriminate between quantities moving out of or into the solution domain.

There exist several definitions for hyperbolicity. Here, two definitions are given which are useful when discussing the Euler Equations. A first-order system,

$$\frac{\partial}{\partial t}\mathbf{u} = \mathbf{P}_1(\frac{\partial}{\partial x_j})\mathbf{u} = \sum_{j=1}^{3} \mathbf{B}_j D_j \mathbf{u} \tag{2.2.17}$$

is called

- *weakly hyperbolic* if for all $\mathbf{k} \in \mathbf{R}^3$ the *eigenvalues* of the symbol $\mathbf{P}_1(ik_j)$ are purely imaginary,
- *symmetric hyperbolic* if $\mathbf{B}_j = \mathbf{B}_j^T$ for $j = 1, 2, 3$.

One can show that, in general, initial-boundary value problems of weakly hyperbolic systems can become ill-posed if lower-order terms are present.

Symmetric hyperbolic systems have a symbol of the form

$$\mathbf{P}_1(ik_j) = i\sum_{j=1}^{3} k_j \mathbf{B}_j \tag{2.2.18}$$

where the sum, $\sum_{j=1}^{3} k_j \mathbf{B}_j$, is a Hermitian matrix. Therefore, $\mathbf{P}_1(ik_j)$ can be diagonalized by a unitary transformation which ultimately allows to formulate well-posed initial-boundary-value problems under very restrictive conditions. Specifically, the estimation of the boundary term in eq. (2.2.15), $\int_0^t \|\mathbf{u}(\tau)\|_\Gamma^2 \, d\tau$, causes difficulties in more than one space dimension. Symmetric hyperbolic systems are of interest because the Euler Equations can be cast into this form.

As already mentioned it is not generally possible to formulate well-posed problems for general hyperbolic systems in more than one space dimension, and in principle it would be necessary to show well-posedness for each special case. As this is not easily possible one relies on numerical experimentation: with boundary conditions considered reasonable one tries to solve the equations numerically and decides if the result is physically acceptable.

The formulation of boundary conditions is best done using the concept of *characteristic variables* which is illustrated for one space dimension and then generalized to higher dimensions.

It is sufficient to consider the half-space problem, $x \in [0, \infty)$, since it can be shown that the results obtained there can be generalized to more complicated solution domains. Consider the *hyperbolic system*

$$\mathbf{u}_t = \mathbf{B}(x,t)\mathbf{u}_x \,, \quad x \in [0,\infty) \,, \quad \mathbf{u}(x,0) = \mathbf{f}(x) \,. \tag{2.2.19}$$

From hyberbolicity follows that the eigenvalues, $\lambda_j(x,t)$, of $\mathbf{B}(x,t)$ are real, i.e. there exists a smooth transformation $\mathbf{S}(x,t)$:

$$\mathbf{S}^{-1}\mathbf{B}\mathbf{S} = \mathbf{\Lambda}(x,t) = \left\{ \begin{matrix} \lambda_1(x,t) & & 0 \\ \cdot & \cdot & \cdot \\ 0 & & \lambda_n(x,t) \end{matrix} \right\} . \tag{2.2.20}$$

Further it is assumed that the eigenvalues do not change sign as a function of time on the boundary $x = 0$. Introducing *characteristic variables*,

$$\hat{\mathbf{u}} = \mathbf{S}^{-1}\mathbf{u} \,, \tag{2.2.21}$$

leads to a *decoupled system*:

$$\hat{\mathbf{u}}_{j,t} = \lambda_j(x,t)\hat{\mathbf{u}}_{j,x} \,, \quad j = 1, \ldots, n \,. \tag{2.2.22}$$

Arranging $\mathbf{\Lambda}(x,t)$ on the boundary as

$$\mathbf{\Lambda}(0,t) = \begin{Bmatrix} \mathbf{\Lambda}_+ & & 0 \\ & \mathbf{\Lambda}_0 & \\ 0 & & \mathbf{\Lambda}_- \end{Bmatrix} \tag{2.2.23}$$

where $\mathbf{\Lambda}_+$ contains the positive eigenvalues, $\mathbf{\Lambda}_0$ the zero eigenvalues, and $\mathbf{\Lambda}_-$ the negative eigenvalues. The components of $\hat{\mathbf{u}}$ are arranged accordingly, $\hat{\mathbf{u}}_+$, $\hat{\mathbf{u}}_0$, $\hat{\mathbf{u}}_-$. One then can prove the following theorem:

The problem (2.2.19) together with the boundary conditions

$$\hat{\mathbf{u}}_-(0,t) = \mathbf{g}(t) \tag{2.2.24}$$

is well posed.

The theorem states that one has to prescribe boundary conditions only for characteristic variables entering the solution domain ($\lambda_j < 0$). Boundary values of characteristic variables leaving the solution domain ($\lambda_j > 0$) cannot be prescribed. They are determined from the inner solution. The characteristic variables belonging to *zero eigenvalues* are determined from the initial conditions, $\hat{\mathbf{u}}_0(x,t) = (\mathbf{S}^{-1}\mathbf{f}(x))_0$, and boundary conditions are neither necessary nor allowed. (To avoid discontinuities propagating along the characteristics into the solution domain, one requires compatibility between initial and boundary conditions, usually $(\mathbf{S}^{-1}\mathbf{f}(x=0))_0 = \mathbf{g}(t=0)$).

The theorem can be extended to systems of the form

$$\mathbf{u}_t = \mathbf{B}(x,t)\mathbf{u}_x + \mathbf{C}(x,t)\mathbf{u} + \mathbf{F}(x,t) \tag{2.2.25}$$

with the boundary conditions specified as in (2.2.24).

The formulation of a well-posed problem in higher dimensions proceeds along the same lines as in the one-dimensional case. Assume the hyperbolic system:

$$\begin{gathered} \mathbf{u}_t = \sum_{i=1}^{3} \mathbf{B}_i(\mathbf{x},t)D_i\mathbf{u} + \mathbf{C}(\mathbf{x},t)\mathbf{u} + \mathbf{F}(\mathbf{x},t) \,, \qquad \mathbf{x} \in \mathbf{R}^+ \times \mathbf{R}^2 \\ \mathbf{u}(\mathbf{x},0) = \mathbf{f}(\mathbf{x}) \,, \end{gathered} \tag{2.2.26}$$

where periodic boundary conditions are prescribed in y and z. The fundamental difference to the one-dimensional case is the requirement that the system (2.2.26) has to be symmetric, i.e. $\mathbf{B}_i(\mathbf{x},t) = \mathbf{B}_i^T(\mathbf{x},t)$. The symmetry of the $\mathbf{B}_i$'s is necessary in order to make boundary contributions in eq. (2.2.15) vanish which otherwise could not be estimated.

For simplicity, it is assumed that $\mathbf{B}_1$ is already in diagonal form and ordered as before:

$$\mathbf{B}(0,t) = \mathbf{\Lambda}(0,t) = \left\{ \begin{matrix} \mathbf{\Lambda}_+ & & 0 \\ & \mathbf{\Lambda}_0 & \\ 0 & & \mathbf{\Lambda}_- \end{matrix} \right\} \tag{2.2.27}$$

and the solution vector is also subdivided accordingly. One then has the theorem:

The problem (2.2.26) is well-posed with the boundary conditions

$$\mathbf{u}_-(0,y,z,t) = \mathbf{g}(y,z,t)\ , \tag{2.2.28}$$

i.e. only incoming characteristic variables require boundary conditions.

In applications the equation system is rarely in characteristic form. The problem then is to identify which variable combinations need boundary conditions, and how the boundary values, $\mathbf{g}(y,z,t)$ have to be chosen in order to describe the physical situation one wants to model. Or, if characteristic variables are not determined, which boundary conditions applied to the noncharacteristic variables are equivalent to (2.2.28). Some of the difficulties arising when using characteristic variables are discussed in the next chapter.

As stated above, the boundary values of characteristic variables leaving the solution domain are to be determined from information inside. For open boundaries which are totally artificial it is difficult to construct a form of boundary conditions which are transpararent to disturbances moving from the inside towards the boundary despite the fact that reflecting boundary conditions lead to well-posed problems but do not represent the physical problem.

Mixed Systems. In contrast to the Euler Equations the Navier-Stokes Equations cannot be classified according to the scheme given above because the continuity equation does not contain second derivatives. One has rather a *mixed system* of parabolic and hyperbolic equations. For equations of the Navier-Stokes type one can show that well-posedness is not destroyed by lower-order coupling terms if boundary conditions are formulated for the parabolic and hyperbolic components separately according to the techniques discussed in this chapter.

3. Boundary Conditions

When treating boundary conditions two aspects have to be considered: first, the formal statement of the boundary conditions, and second their discrete representations. As the discrete formulations are intimately related to the type of grid chosen (e.g. a staggered grid or a regular grid), to the specific form of the equations (e.g. flux-form or transport form), and to the integration algorithm chosen, the following notes give a discussion on the formal aspects with hints what parts need special attention, and with less emphasis on the specific implementation.

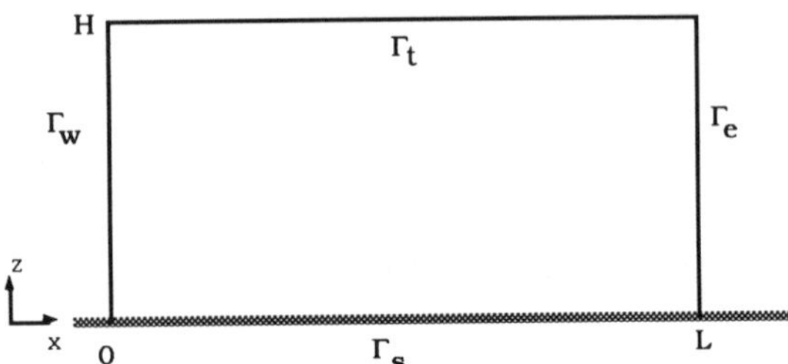

Figure 2.3.1: Solution domain for a limited area model. The boundaries Γ are labelled according to s = surface, w = west, e = east, t = top.

In limited-area atmospheric models the surface is the only physical bounary of the solution domain. All other boundaries are purely computational, and it is up to the ingenuity of the modeller to device boundary conditions which allow to shift the location of the open boundaries without affecting the solution in the area of interest.

The atmosphere behaves more or less like stratified fluid in which turbulence is mostly suppressed, and wave-like excitations can traverse the solution domain without appreciable damping. The Euler Equations are therefore in many cases a good approximation, and boundary conditions are needed for hyperbolic systems. However, wave propagation in the system is strongly anisotropic. This in turn is taken as justification to treat the boundary conditions for the top, Γ_t, and for the sides, Γ_w and Γ_e, separately.

During strong solar irradiation the atmosphere in the lower troposphere largely behaves as a turbulent flow. This case is adequately described by the parabolic version of the equations for which the formulation of boundary conditions is fairly settled. Based on this knowledge, a boundary treatment of the Euler Equations is justified where, close to the open boundaries, the type of the equations is changed to parabolic in order to avoid some of the difficulties with hyperbolic systems.

The Incompressible Navier-Stokes Equations

The incompressible N-S Equations (2.1.30) (for simplicity without Coriolis force),

$$\begin{aligned} \rho_0 \frac{d}{dt}\mathbf{u} + \nabla p' - \mu_m \Delta \mathbf{u} &= -\frac{\rho_0}{T_0}\mathbf{g}T' \\ \rho_0 \frac{d}{dt} T &= \frac{\mu_T}{c_V}\Delta T \\ \nabla \cdot \mathbf{u} &= 0 \end{aligned} \tag{2.3.1}$$

($T = T_0 + T'$) are of parabolic type, i.e. μ_m and μ_T are large enough to sufficiently damp wave-like excitations such that reflections at the boundaries can be neglected, and boundary conditions for the dynamical variables can be chosen according to (2.2.13).

Since the continuity equation is reduced to the incomressibility constraint pressure has to be adapted accordingly. Applying the divergence operator to the first equation in (2.3.1) yields a Poisson Equation for pressure:

$$\Delta p' = -\rho_0 \nabla \cdot [(\mathbf{u} \cdot \nabla)\mathbf{u}] - \frac{\rho_0}{T_0} \nabla \cdot (\mathbf{g}T') \tag{2.3.2}$$

It is generally accepted that no boundary conditions have to be given for the pressure equation (Gresho 1987, 1991). They are rather determined from values of velocity and temperature on the boundary.

Surface. When the lower boundary represents a surface (impenetrable wall) the velocities have to fulfill Dirichlet conditions, and for temperature mixed conditions may be prescribed:

$$u(x,0,t) = w(x,0,t) = 0\ , \quad cT_z(x,0,t) + dT(x,0,t) = T_s(x,t)\ , \quad x \in \Gamma_s\ , t > 0 \tag{2.3.3}$$

For applications in which the boundary layer is not resolved mixed boundary conditions for the tangential velocity, **u**, are appropriate:

$$au_z(x,0,t) + bu(x,0,t) = u_0(x,t)\ . \tag{2.3.4}$$

Inflow. At inflow the velocity vector points into the solution domain, and profiles of velocity and temperature can be given. For example, let Γ_w be the inflow boundary. Then

$$u(0,z,t) = U(z,t)\ , \quad w(0,z,t) = 0\ , \quad T(0,z,t) = T_B(z,t)\ , \quad z \in \Gamma_w\ . \tag{2.3.5}$$

Outflow. On the outflow boundary, Γ_e, the velocity vector points out of the solution domain. One may, for example, set the second derivatives to zero. This amounts to linearly extrapolating the variables from inside:

$$u_{xx}(L,z,t) = w_{xx}(L,z,t) = T_{xx}(L,z,t) = 0\ , \quad z \in \Gamma_e\ . \tag{2.3.6}$$

Top At the top of the solution domain a convenient choice is

$$u(x,H,t) = U(z,t)\ , \quad w(x,H,t) = 0\ , \quad T(x,H,t) = T_B(H,t)\ , \quad x \in \Gamma_t\ . \tag{2.3.7}$$

The given boundary conditions are not independent since from eq. (2.3.1)

$$\int_V \nabla \cdot \mathbf{u} d\mathbf{r} = \oint_\Gamma (\mathbf{u} \cdot d\mathbf{s}) = 0\ , \tag{2.3.8}$$

i.e. there must be a mass balance not only in the interior but also on the boundary. In addition, the initial conditions must be mass-balanced. The simplest way to enforce the constraint (2.3.8) is to force also the gradients of velocities on the boundary to zero.

Pressure. As already stated the *pressure* field in incompressible flow must be adapted in such a way that the divergence condition is fulfilled everywhere in the solution domain. An elegant solution to this problem is the pressure correction method (Harlow and Welch, 1965). For simplicity, an explicit time integration method is assumed, and buoyancy is neglected:

a) Integrate the velocity field at time step n, $\mathbf{u}^n$, to an intermediate time '*' leaving out the pressure gradient ($\tau = \Delta t$):

$$\mathbf{u}^* = \mathbf{u}^n - \tau[(\mathbf{u}^n \cdot \nabla)\mathbf{u}^n + \nu_m \Delta \mathbf{u}^n] \qquad (2.3.9)$$

With this definition the velocity at time $n+1$ differs from $\mathbf{u}^*$ by the pressure gradient:

$$\mathbf{u}^{n+1} = \mathbf{u}^* - \tau \frac{1}{\rho_0} \nabla p' \qquad (2.3.10)$$

Applying the divergence operator to (2.3.10) and requiring a divergence-free velocity field at time $n+1$ yields a Poisson equation for pressure. Therefore the second step is:

b) solve the Poisson equation

$$\frac{1}{\tau} \nabla \cdot (\rho_0 \mathbf{u}^*) = \Delta p' \,. \qquad (2.3.11)$$

c) Update the intermediate velocity, $\mathbf{u}^*$, to the final result:

$$\mathbf{u}^{n+1} = \mathbf{u}^* - \tau \frac{1}{\rho_0} \nabla p' \,. \qquad (2.3.12)$$

Equation (2.3.11) is *elliptic.* It therefore needs either the boundary values and/or the normal derivatives. As only the normal derivative of pressure can be obtained from the momentum equations the pressure is determined up to a constant.

Surface. As $w = 0$ at the lower and upper boundaries on has

$$p'_z(x, z, t) = 0 \,, \quad z = 0, H \,, \quad x \in \Gamma_s, \Gamma_t \,. \qquad (2.3.13)$$

Open Boundaries. Assuming the normal gradients of the velocity field to be zero on the boundaries the pressure derivatives are

$$u_t(x, z, t) = -\frac{1}{\rho_0} p'_x(x, z, t) \,, \quad x = 0, L \,, \quad z \in \Gamma_w, \Gamma_e \,. \qquad (2.3.14)$$

The Euler Equations

Discarding the diffusion terms in (2.1.26) yields the Euler Equations for incompressible stratified flow

$$\overline{\rho} \frac{d}{dt} \mathbf{u} + \nabla p' = -\frac{\overline{\rho}}{\overline{\theta}} \mathbf{g} \theta' \,, \quad \nabla \cdot (\overline{\rho} \mathbf{u}) = 0 \,, \quad \overline{\rho} \frac{d}{dt} \theta = 0 \qquad (2.3.15)$$

where $\theta = \overline{\theta} + \theta'$. In the following only problems arising from open boundaries are discussed as boundary conditions at solid surfaces are easily realised (zero gradient for tangential velocities, and zero for normal velocities).

Open boundaries arise from an artificial limitation of the solution domain due to restricted computational resources. The central problem is then how to formulate open boundary conditions such that their influence is not felt. In principle, one should be able to shift the open boundaries without affecting the solution. More specifically, the boundary conditions should be transparent to perturbations moving from inside the solution domain towards the boundaries. Reflected perturbations should be small, i.e. the boundary conditions should be absorbing for reflected waves.

There are essentially three approaches to tackle the open boundary problem for the Euler equations. The so-called *sponge-layer* technique consists in changing the type of the equations from hyperbolic to parabolic in the vicinity of the boundary. Then the knowledge on boundary conditions from parabolic equations can be taken over.

In the second approach one approximates the equations on the boundary by simpler hyperbolic equations which are then used to determine the boundary values. The coefficients of the simplified equations are determined numerically.

In the third approach the *characteristic variables* are determined from the linearized equations, and approximate boundary conditions are constructed. Though the latter technique is mathematically attractive it is difficult to apply in higher dimensions. Especially stratification causes considerable difficulties to find numerically useful formulas.

Sponge-Layers. Suppose boundary conditions for the equations (2.3.15) are sought. Then, one introduces a *sponge layer* close to the boundaries (Figure 2.3.2) in which diffusion terms are added to the dynamical equations increasing from zero to a size such that the Reynolds Number is considerably reduced. For example, the equation for the u-velocity close to the left boundary, Γ_w, would be:

$$\overline{\rho}\frac{d}{dt}u + \frac{\partial}{\partial x}p' = \overline{\rho}\frac{\partial}{\partial x}K_u(x)\frac{\partial}{\partial x}u\,. \tag{2.3.16}$$

The spacial dependence of K_u has to be smooth enough such that no artificial boundary layer

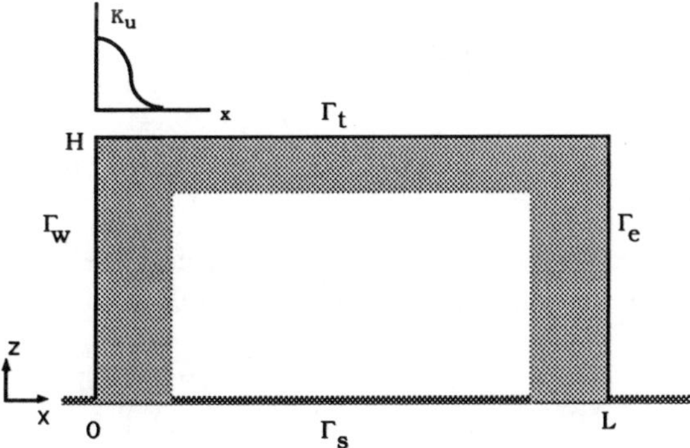

Figure 2.3.2: Sponge layer in the boundary regions of the solution domain.

is generated, and waves moving from the inside towards the boundary are only marginally reflected at the inhomogeneity of K_u. On the other hand economy requires not to waste a too big sponge area. One is therefore forced to experimentation. A boundary treatment of the form (2.3.16) has been discussed, e.g., by Burridge (1975).

Another way of changing the equations is to introduce a *relaxation term* forcing the variable towards the desired boundary value $\tilde{u}$:

$$\overline{\rho}\frac{d}{dt}u + \frac{\partial}{\partial x}p' = -\alpha_u(x)\overline{\rho}(u - \tilde{u}) \qquad (2.3.17)$$

Yet another method is the combination of the last two techniques (see e.g. Tatsumi, 1980):

$$\overline{\rho}\frac{d}{dt}u + \frac{\partial}{\partial x}p' = -\alpha_u(x)\overline{\rho}(u - \tilde{u}) + \frac{\partial}{\partial x}K_u(x)\frac{\partial}{\partial x}u\,. \qquad (2.3.19)$$

A systematic investigation of the properties of the above treatments has been given by Davies (1983).

Wave Excitations. The Euler Equations (2.3.15) support gravity waves. Choosing a stably stratified background state ($\Gamma > 0$),

$$\overline{\theta}(z) = \theta_0 + \Gamma z\,, \qquad \frac{\partial}{\partial z}(\ln\overline{\rho}) = -\beta \simeq 0\,, \qquad (2.3.20)$$

and linearizing equations (2.3.15) around a constant flow ($\mathbf{U} = (U, 0)$),

$$\begin{aligned} u_t + Uu_x &= -\frac{1}{\rho_0}p_x \\ w_t + Uw_x &= -\frac{1}{\rho_0}p_z + \frac{g}{\theta_0}\theta \\ \theta_t + U\theta_x + \Gamma w &= 0 \\ u_x + w_z &= 0\,, \end{aligned} \qquad (2.3.21)$$

one arrives at *dispersion relations* of the form:

$$\omega_1 = k_1 U\,, \qquad \omega_{2,3} = k_1\left(U \pm \frac{N}{\sqrt{k_1^2 + k_2^2}}\right) \qquad (2.3.22)$$

which determines the propagation of a Fourier component, $\exp i(k_1 x + k_2 z - \omega t)$. The *Brunt-Väisälä frequency* is given by $N = \sqrt{\Gamma g}$. As the *phase velocity,*

$$\mathbf{v}_p = \begin{Bmatrix} \omega/k_1 \\ \omega/k_2 \end{Bmatrix} = \begin{Bmatrix} U \pm \frac{N}{|k|} \\ \frac{k_1}{k_2}\left(U \pm \frac{N}{|k|}\right) \end{Bmatrix}, \qquad (2.3.23)$$

is wave-number dependent each Fourier component travels at different speed. From (2.3.23) one can see that for some horizontal wave-numbers, k_1, the phase velocity can even change sign without a change of sign of the wave number. So part of the Fourier components may be incoming waves and part on outgoing. In configuration space this would imply nonlocal boundary conditions. Many of the techniques suggested so far are aimed at mimicking this inherent nonlocality by *local approximations* (a review is given by Givoli, 1991).

For vertically propagating waves ($k_1 = 0$) a slight simplification occurs as the propagation direction of perturbations is determined by the sign of the wavenumber. Upward and downward moving waves can easily be separated.

This anisotropy in wave propagation serves as justification to treat the horizontal and vertical boundaries separately.

Top Boundary To identify the upward moving wave contribution one goes back to equation (2.3.21) and inserts ($\sigma = \{u, w, \theta', p'\}$)

$$\sigma = \hat{\sigma}(z)\exp[i(k_1 x - \omega t)] . \tag{2.3.24}$$

One obtains ($\epsilon = -\omega + k_1 U$):

$$\begin{aligned} i\epsilon\hat{u} &= -i\frac{k_1}{\rho_0}\hat{p} \\ i\epsilon\hat{w} &= -\frac{1}{\rho_0}p_z + \frac{g}{\theta_0}\hat{\theta} \\ i\epsilon\hat{\theta} &= -\Gamma\hat{w} \\ ik_1\hat{u} + \hat{w}_z &= 0 \end{aligned} \tag{2.3.25}$$

Solving for $\hat{w}$ yields a wave equation:

$$\left\{\frac{\partial}{\partial z} + i\frac{k_1}{\epsilon}\sqrt{N^2 - \epsilon^2}\right\}\left\{\frac{\partial}{\partial z} - i\frac{k_1}{\epsilon}\sqrt{N^2 - \epsilon^2}\right\}\hat{w} = 0 \tag{2.3.26}$$

with the general solution

$$\hat{w} = a\exp(i\lambda z) + b\exp(-i\lambda z)\,, \quad \lambda = \frac{k_1}{\epsilon}\sqrt{N^2 - \epsilon^2}\,. \tag{2.3.27}$$

The same form of equation is obtained for the other variables. Together with the factor $exp(-i\omega t)$ (eq. 2.3.24) one identifies the first contribution in (2.3.27) with the wave travelling upward, and the second downward.

The general solution is a superposition of arbitrary shapes travelling with the same speed in opposite directions (Figure 2.3.3),

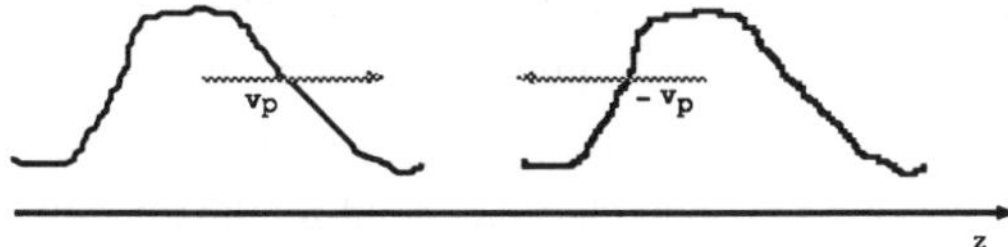

Figure 2.3.3: Solutions to the One-dimensional Wave Equation

and the operator

$$\left\{\frac{\partial}{\partial z} - i\frac{k_1}{\epsilon}\sqrt{N^2 - \epsilon^2}\right\} \tag{2.3.28}$$

transports the upward moving wave out without reflections; i.e. it anihilates downgoing wave conponents. Boundary conditions which use some form of projection to eliminate the *reflected wave* are called *absorbing boundary conditions.*

Applying the operator (2.3.28) at the upper boundary to the individual variables would therefore be an optimal boundary condition. However, ϵ contains the frequency. The whole time history of the field would be necessary to apply (2.3.28). A simplification which seems to be satisfactory is to neglect the ω-dependence and use instead (Klemp and Durran, 1983):

$$\left\{\frac{\partial}{\partial z} - i\frac{k_1}{k_1 U}\sqrt{N^2 - (k_1 U)^2}\right\}\hat{w} = 0\,. \tag{2.3.29}$$

The upper boundary treatment could then proceed in the following steps:

a) Calculate the horizontal Fourier Transforms of the variables (u, w, θ) at the upper boundary.

b) Project on the up-going wave components by applying eq. (2.3.29).

c) Transform back to coordinate space.

Boundary Values for Pressure. As stated before, the pressure boundary has to be obtained from the dynamical equations. Solving the system for $\hat{p}$ gives:

$$-i\frac{k_1^2}{\epsilon}\hat{p} = -\rho_0 \hat{w}_z\,. \tag{2.3.30}$$

The velocity derivative is eliminated by the radiation condition (2.3.24) resulting in the desired boundary condition for pressure:

$$\hat{p} = \frac{\sqrt{n^2 - \epsilon^2}}{|k_1|}\hat{w}\,. \tag{2.3.31}$$

Again, the simplest application of (3.29) is to set $\omega = 0$. Once the Fourier Components for w are available the transformation back to coordinate space of equation (2.3.30) yields the desired pressure boundary values.

Horizontal Boundary Conditions In most applications the mean velocity, U, is nonzero, and the treatment of the horizontal boundaries cannot be shaped along the procedure for the vertical boundary.

An elementary but fairly successful way out was put forward by Orlanski (1976) whose original suggestion to treat orthogonally incident waves is widely used, and later improvements allowing for inclined incident waves were put forward (e.g. Raymond and Kuo, 1984).

As the hydrodynamical equations are nonlinear and the linear analysis need not be reliable, and as a quantity ϕ approaching the boundary Γ_w head-on the operator

$$(\frac{\partial}{\partial t} - c^*\frac{\partial}{\partial x})\phi = 0 \tag{2.3.32}$$

completely absorbs this wave. The unknown phase velocity should be determined from the behaviour of the flow adjacent to the boundary and inside the solution domain. Therefore, Eq. (2.3.32) can be used to determine the local phase velocity from the flow itself:

$$c^* = \frac{\phi_t}{\phi_x} . \tag{2.3.33}$$

If only grid point values already calculated are used, equation (2.3.33) is well determined. Then, depending on the sign of c^*, either an upwind step is performed transporting ϕ out or external boundary conditions, e.g. $\phi^{n+1} = \phi^n$, are applied.

As this approach is highly experimental some experience is needed to generate good results. For example, using a slightly higher phase velocity than is numerically determined shows less reflections.

The distinct feature of this boundary treatment is its inherent nonlinearity as no linearization of the system is needed.

Characteristics.

Using characteristic information to construct boundary conditions is attractive insofar as dynamical information is used. Techniques based on this approach have been very successful for scalar equations. However, treating systems containing waves which hit the boundary at non-zero angle is difficult. An outline of the approach is given below.

Based on the discussion at the end of section **2** one can try to use characteristic information to treat the open boundaries. The approaches differ from each other by the manner how characteristic information is obtained and what simplifying assumptions are made. The principle is illustrated with the 1-d Shallow Water Equations.

The 1-d Shallow Water Equations. The 1-d Shallow Water Equations on the half-space $[0, \infty)$ are given in the form

$$\frac{\partial}{\partial t}\mathbf{u} = \frac{\partial}{\partial t}\begin{Bmatrix} u \\ h \end{Bmatrix} = \frac{\partial}{\partial x}\begin{Bmatrix} -Uu - gh \\ -Hu - Uh \end{Bmatrix} = \begin{Bmatrix} -U & -g \\ -H & -U \end{Bmatrix}\frac{\partial}{\partial x}\begin{Bmatrix} u \\ h \end{Bmatrix} = \mathbf{B}\frac{\partial}{\partial x}\mathbf{u} . \tag{2.3.34}$$

Performing an eigenvalue decomposition of $\mathbf{B}$ as discussed at the end of section **2** yields:

$$\begin{matrix} \lambda_1 = -c + U \\ \lambda_2 = c + U \end{matrix} , \quad c = \sqrt{gH} , \quad \mathbf{e}_1 = \begin{Bmatrix} \sqrt{g/H} \\ 1 \end{Bmatrix} , \quad \mathbf{e}_2 = \begin{Bmatrix} -\sqrt{g/H} \\ 1 \end{Bmatrix} . \tag{2.3.35}$$

For subcritical flow, $0 < U < c$ ($c \simeq 90$ m/s for $H = 8$ km), only the characteristic variable $\mathbf{e}_1$ enters the solution domain at $x = 0$. One has to prescribe boundary conditions ("o" = outer values):

$$\sqrt{\frac{g}{H}}u_o + h_o = b_0 . \tag{2.3.36}$$

The value of the characteristic variable $\mathbf{e}_2 = -\sqrt{g/H}u + h$ on the boundary has to be determined from values inside the solution domain, $x > 0$. A discrete approximation to

$$\frac{\partial}{\partial t}\mathbf{e}_2 - \lambda_2\mathbf{I}\frac{\partial}{\partial x}\mathbf{e}_2 = 0 \tag{2.3.37}$$

serves this purpose (e.g. an upwind approximation) yielding the value $b_i = -\sqrt{g/H}u_o + h_o$ ("i" = inner values) determined from inside. Then, the values of the dynamical variables, u_o and h_o can be determined. For example, with u_o prescribed and b_i determined from inside one has $h_o = \sqrt{g/H}u_o + b_i$.

It should be noted that, depending on the actual prescription of how the boundary values are generated from inner information the stability of the whole scheme has to be checked, either by investigation of the Fourier modes or by numerical experimentation (see e.g. Foreman, 1986).

A nonlinear approach has been suggested by Hedström (1978).

The simplicity of the procedure just discusssed is due to the lack of dispersion in the system. Dispersion, however, is always present in multidimensional flow whenever characteristic variables hit the boundary of the solution domain at inclined angles (see below). It is therefore desirable to find approximate boundary conditions for these cases.

Exact Absorbing Boundary Conditions. As an example for a system with dispersion (which seems to be the only simple one) take the wave equation,

$$\left[\frac{\partial^2}{\partial t^2} + c^2\left(\frac{\partial^2}{\partial x^2} + \frac{\partial^2}{\partial z^2}\right)\right] u(x, z, t) = 0 \tag{2.3.38}$$

in the half space $x > 0$ with the dispersion relation

$$\omega_{1,2} = \pm c\sqrt{k_1^2 + k_2^2}\,. \tag{2.3.39}$$

The aim is to construct a boundary condition at the artificial boundary $x = 0$ which is transparent to waves moving from the area $x > 0$ towards $x = 0$ (Figure 2.3.4). Reflected waves should be "absorbed" to prevent them from returning into solution domain and thus contaminating the solution.

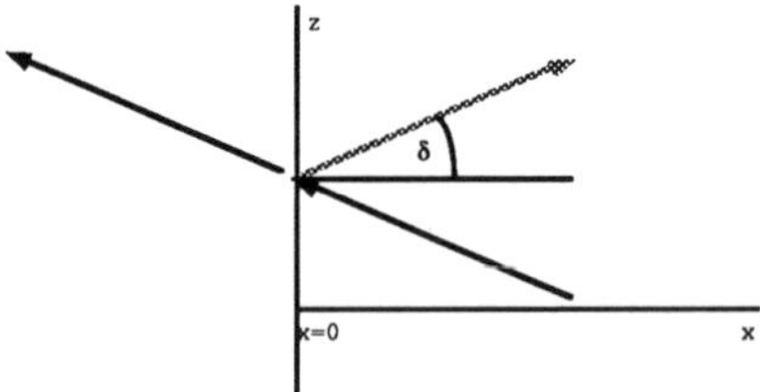

Figure 2.3.4: Wavenumber vector hitting the artificial boundary at $x = 0$.

Solving equation (2.3.39) for k_1, one can write a wave component travelling to the left as

$$u = \exp i\left[\sqrt{\left(\frac{\omega}{c}\right)^2 - k_2^2}\,x + k_2 z + \omega t\right], \quad \omega > 0\,, \quad \left(\frac{\omega}{c}\right)^2 > k_2^2\,. \tag{2.3.40}$$

The operator anihilating such a component is

$$\left[\frac{\partial}{\partial x} - i\sqrt{\left(\frac{\omega}{c}\right)^2 - k_2^2}\right] u(x=0) = 0\,. \tag{2.3.41}$$

An arbitrary wave packet being absorbed at $x = 0$ has the form

$$u(x,z,t) = \int d\omega \int dk_2 \exp i\left[\sqrt{\left(\frac{\omega}{c}\right)^2 - k_2^2}\, x + k_2 z + \omega t\right] \hat{u}(k_2,\omega) \tag{2.3.42}$$

and obeys the equation (2.3.41):

$$\frac{\partial}{\partial x}u(x,y,t) - \int d\omega \int dk_2 i\sqrt{\left(\frac{\omega}{c}\right)^2 - k_2^2} \exp i\left[\sqrt{\left(\frac{\omega}{c}\right)^2 - k_2^2}\, x + k_2 z + \omega t\right] \hat{u}(k_2,\omega)\,. \tag{2.3.43}$$

Nirenberg (1973) has shown that so called pseudo-differential operators can be defined through their Fourier Transforms,

$$\begin{aligned} &-\sqrt{\left(\frac{\partial^2}{\partial t^2} - \frac{\partial^2}{\partial z^2}\right)}\, u(x=0) \\ &=_{def} \int d\omega \int dk_2 i\sqrt{\left(\frac{\omega}{c}\right)^2 - k_2^2} \exp i\left[\sqrt{\left(\frac{\omega}{c}\right)^2 - k_2^2}\, x + k_2 z + \omega t\right] \hat{u}(k_2,\omega)\,, \end{aligned} \tag{2.3.44}$$

as asymptotic series (to keep the presentation free from technical burden factors are left out which are needed for convergence of the integral when the angle of incidence approaches ninety degrees). The absorbing boundary condition can then be formally written as:

$$\left(\frac{\partial}{\partial x} - \sqrt{\frac{1}{c^2}\frac{\partial^2}{\partial t^2} - \frac{\partial^2}{\partial z^2}}\right) u(x=0,z,t) = 0\,. \tag{2.3.45}$$

It is justified to make expansions algebraically in Fourier space and then switch to differential operators:

$$-i\omega \quad \leftrightarrow \quad \frac{\partial}{\partial t}\,, \qquad ik_2 \quad \leftrightarrow \quad \frac{\partial}{\partial z}\,. \tag{2.3.46}$$

With this methodology one can derive a hierarchy of local boundary condition which lead to well-posed problems.

1. Approximation. Expanding the algebraic counterpart of (2.3.12) up to first order yields

$$\frac{\partial}{\partial x} - i\sqrt{\left(\frac{\omega}{c}\right)^2 - k_2^2} \simeq \frac{\partial}{\partial x} - i\frac{\omega}{c}\sqrt{1 - \left(\frac{ck_2}{\omega}\right)} \simeq \frac{\partial}{\partial x} - i\frac{\omega}{c}\,. \tag{2.3.47}$$

Going back to differential operators:

$$\left(\frac{\partial}{\partial x} - \frac{1}{c}\frac{\partial}{\partial t}\right) u(x=0) = 0 \quad \rightarrow \quad \left(\frac{\partial}{\partial t} - c\frac{\partial}{\partial x}\right) u(x=0) = 0\,. \tag{2.3.48}$$

2. Approximation (1. Padé Approximant). The next higher coefficient of the square root in (2.3.46),

$$\sqrt{1-\left(\frac{ck_2}{\omega}\right)} \simeq 1-\frac{1}{2}\left(\frac{ck_2}{\omega}\right), \tag{2.3.49}$$

leads to

$$\left[i\omega\frac{\partial}{\partial x}+\left(\omega^2-\frac{1}{2}c^2k_2^2\right)\right] \quad\rightarrow\quad \left(\frac{\partial}{\partial t}\frac{\partial}{\partial x}-\frac{\partial^2}{\partial t^2}+\frac{c^2}{2}\frac{\partial^2}{\partial z^2}\right)u(x=0)=0\,. \tag{2.3.50}$$

Both approximations are stable. One can show that the higher Taylor expansions lead to unstable boundary conditions, which again stresses the need to check stability.

Reflection coefficients can be calculated. Taking $\omega/c = 1$, then $\sqrt{1-k_2^2} = \cos\delta$ and $k_2 = \sin\delta$. Incident and reflected waves are then $a\exp(i\cos\delta\ x + i\sin\delta\ z + i\omega t)$ and $b\exp(-i\cos\delta\ x + i\sin\delta\ z + i\omega t)$, respectively. The amplitudes are then

$$\frac{b}{a}=\left|\frac{\cos\delta-1}{\cos\delta+1}\right|, \qquad \frac{b}{a}=\left|\frac{\cos\delta-1}{\cos\delta+1}\right|^2 \tag{2.3.51}$$

for the first and second approximation. The higher approximation extends the cone within which absorption is high.

The 2-d Shallow Water Equations. The extension to higher space dimensions is nontrivial due to 'geometrical' (and possibly also dynamical) dispersion. The open boundaries proposed Engquist and Majda (1977) show remarkable permeability for the 2-d Shallow Water Equations (see also Durran and Yang, 1993). Recently, this approach has also been applied to the treatment of open boundaries of the nonlinear 2-d Euler Equations (Kröner, 1991) . So this method is likely to improve the boundary treatment of atmospheric models in the horizontal (x-y) direction.

To present the technique while keeping the formulas small equations (2.3.34) are generalized to two dimensions with a mean flow in x-direction. After rescaling the velocities: $u \rightarrow \sqrt{H/g}\,u$ and $v \rightarrow \sqrt{H/g}\,v$, the Shallow Water Equations are ($c = \sqrt{gH}$):

$$\frac{\partial}{\partial t}\mathbf{u} = \mathbf{B}_1\frac{\partial}{\partial x}\mathbf{u} + \mathbf{B}_2\frac{\partial}{\partial y}\mathbf{u}\,, \qquad \mathbf{u} = \{u, v, h\}^T \tag{2.3.52}$$

with

$$\mathbf{B}_1 = \begin{Bmatrix} -U & 0 & -c \\ 0 & -U & 0 \\ -c & 0 & -U \end{Bmatrix}, \quad \mathbf{B}_2 = \begin{Bmatrix} 0 & 0 & 0 \\ 0 & 0 & -c \\ 0 & -c & 0 \end{Bmatrix}. \tag{2.3.53}$$

The solution domain is assumed to be the right half-plane, $x \in [0, \infty)$ (as both of the matrices, $\mathbf{B}_1$ and $\mathbf{B}_2$, are symmetric a general solution domain can be mapped on the half space).

To find the characteristic variables crossing the line $x = 0$ the eigenvalues and eigenvectors of $\mathbf{B}_1$ are determined. With the transformation matrix

$$\mathbf{U} = \frac{1}{\sqrt{2}}\begin{Bmatrix} 1 & 0 & 1 \\ 0 & \sqrt{2} & 0 \\ -1 & 0 & 1 \end{Bmatrix} \quad \mathbf{U}^{-1} = \frac{1}{\sqrt{2}}\begin{Bmatrix} 1 & 0 & -1 \\ 0 & \sqrt{2} & 0 \\ 1 & 0 & 1 \end{Bmatrix} \tag{2.3.54}$$

equation (2.3.52) is solved for the x-derivative:

$$\left\{\frac{\partial}{\partial x}\hat{\mathbf{u}} - \mathbf{\Lambda}^{-1}\frac{\partial}{\partial t}\hat{\mathbf{u}} + \mathbf{\Lambda}^{-1}\mathbf{U}^{-1}\mathbf{B}_2\mathbf{U}\frac{\partial}{\partial y}\hat{\mathbf{u}}\right\} = 0 \tag{2.3.55}$$

where $\mathbf{\Lambda}^{-1} = [\mathbf{U}^{-1}\mathbf{B}_1\mathbf{U}]^{-1}$ and $\hat{\mathbf{u}} = \mathbf{U}^{-1}\mathbf{u}$.

In order to obtain the ingoing and outgoing waves, equation (2.3.55) is Fourier-transformed with respect to t and y ($\mathbf{w}(x,\omega,k) = FT[\hat{\mathbf{u}}(x,t,y)]$) resulting in:

$$\left\{\frac{\partial}{\partial x} + i[\omega\mathbf{\Lambda}^{-1} + k\mathbf{\Lambda}^{-1}\hat{\mathbf{B}}_2]\right\}\mathbf{w}(x,\omega,k) = \left\{\frac{\partial}{\partial x} + i\mathbf{M}(\omega,k)\right\}\mathbf{w}(x,\omega,k) = 0\,. \tag{2.3.56}$$

The projection of (2.3.56) onto waves moving to the left is then the accurate absorbing boundary condition at $x = 0$. This is achieved by transforming $\mathbf{M}$ to diagonal form using the matrix $\mathbf{V}$:

$$\left\{\frac{\partial}{\partial x} + i\left\{\begin{matrix}\mu_1 & 0 & 0\\ 0 & \mu_2 & 0\\ 0 & 0 & \mu_3\end{matrix}\right\}\right\}\hat{\mathbf{w}}(x,\omega,k) = 0\,, \qquad \hat{\mathbf{w}} = \mathbf{V}^{-1}\mathbf{w}\,. \tag{2.3.57}$$

From equation (2.3.57) follows that all wave components with negative μ are absorbed at $x = 0$. So defining the operator π as projecting onto the components with negative eigenvalues, μ, the accurate absorbing boundary condition can be written as

$$\left\{\frac{\partial}{\partial x} + i\left\{\begin{matrix}\mu_1 & 0 & 0\\ 0 & \mu_2 & 0\\ 0 & 0 & \mu_3\end{matrix}\right\}\right\}\pi\mathbf{V}^{-1}\mathbf{w} = 0 \tag{2.3.58}$$

from which follows as simplest boundary condition

$$\pi\mathbf{V}^{-1}\mathbf{w} = 0\,. \tag{2.3.59}$$

Interpreting $\pi\mathbf{V}^{-1}$ as pseudo-differential operator defined through it's Fourier Transform,

$$\mathbf{P}(\frac{\partial}{\partial t},\frac{\partial}{\partial y})\hat{\mathbf{u}}(x,t,y)\Big|_{x=0} =_{def} \int d\omega \int dk e^{i\omega t - iky}\pi\mathbf{V}^{-1}(\omega,k)\mathbf{w}(x,\omega,k)\Big|_{x=0}, \tag{2.3.60}$$

one can obtain approximations through Taylor expansions.

After some lengthy calculation one finds the eigenvalues, μ_i to be

$$\mu_1 = -\frac{\omega}{U}, \quad \mu_{2,3} = \frac{1}{c^2 - U^2}(\omega U \pm c\,a)\,. \tag{2.3.61}$$

Therefore, the operator π projects onto the first and third components. For the simplest approximation $\omega = 1$, $k = 0$ (zero-angle incidence) $\mathbf{V}^{-1}(3,3) = 1$, and all other elements are zero. The lowest approximation is therefore $\hat{u}_3 = \{\mathbf{U}^{-1}\mathbf{u}\}_3 = 1/\sqrt{2}(h - u) = 0$. The higher approximations can be obtained straightforward but by lengthy calculation.

2.4. Treatment of Topography

When modelling the airflow over topographically structured terrain it is advantageous to use coordinates in which the boundary conditions can be formulated easily. The greater simplicity of the boundary conditions is, however, balanced by the greater complexity of the transformed equation system.

There are essentially three methods used in atmospheric modelling to represent a height-varying lower boundary.

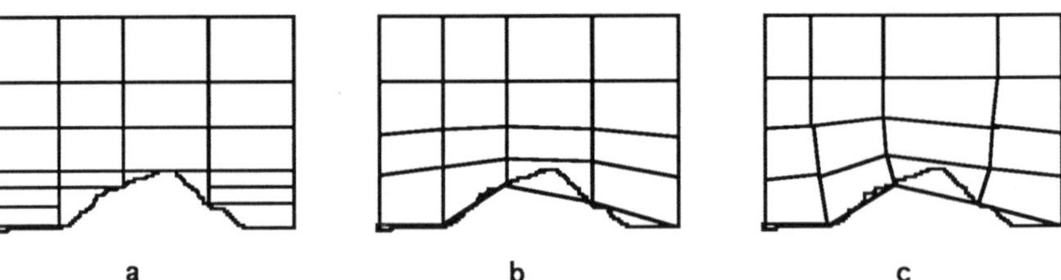

Figure 2.4.1: Different Coordinate Systems

In Figure 2.4.1a the simple approximation of the surface by rectangles is depicted. This representation, though primitive, has its justification when the size of the solution domain is large compared to the boundary layer region. Then, this approximation essentially serves as a means to generate an effective boundary layer which, in its detail, is unphysical but its effect on the flow away from the boundary is realistic.

Figure 2.4.1b is the type of coordinates used in most of the medium to small scale limited-area models (coordinates with pressure or normalized pressure as the vertical coordinate are not considered here). The transformation is chosen in such a way that the horizontal coordinates are unchanged, and only the vertical is mapped such that one coordinate surface coincides with the lower boundary. This choice is decribed below.

Figure 2.4.1c is an example of a so-called Boundary-Fitted Grid (BFG) which is used mainly in aerodynamics and engineering. There, the grid is constructed in such a way that the grid surfaces intersect as orthogonally as possible, and that the individual grid cells have about the same size. These requirements can rarely be fulfilled together. One then chooses either grid refinements or composite grids.

Coordinate Transformations

Though Differential Geometry supplies the elegant apparatus to formulate and investigate the dynamical equations, a more elementary approach is followed here, and only few specific notions are introduced when necessary.

A coordinate transformation is considered as a map from the physical domain into the discrete space of indices (Figure 2.4.2).

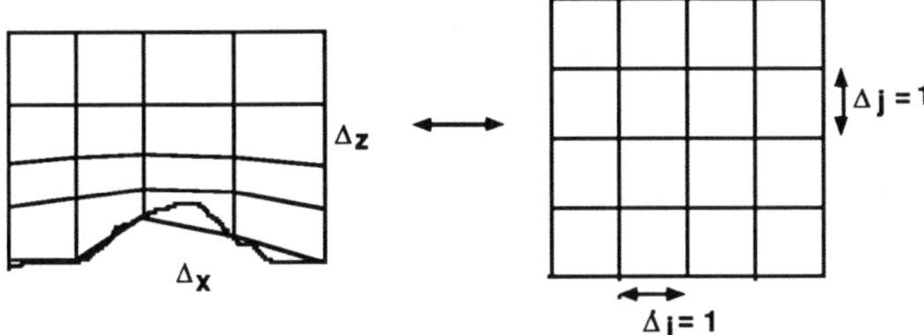

Figure 2.4.2: Coordinate Transformation

It is assumed that, in addition to the Cartesian system with position vector $\mathbf{r}$, a second coordinate system is given with a different representation of the position vector $\mathbf{r}$:

$$\mathbf{r} = x\mathbf{i} + y\mathbf{j} + z\mathbf{k} = \xi\mathbf{e}_1 + \eta\mathbf{e}_2 + \zeta\mathbf{e}_3 \,. \tag{2.4.1}$$

With the left-hand-side of eq. (2.4.1) given one determines the basis vectors in the transformed system (subscripts denote partial differentiation):

$$\begin{aligned}
\frac{\partial \mathbf{r}}{\partial \xi} &= \mathbf{e}_1 = x_\xi \mathbf{i} + y_\xi \mathbf{j} + z_\xi \mathbf{k} \\
\frac{\partial \mathbf{r}}{\partial \eta} &= \mathbf{e}_2 = x_\eta \mathbf{i} + y_\eta \mathbf{j} + z_\eta \mathbf{k} \\
\frac{\partial \mathbf{r}}{\partial \zeta} &= \mathbf{e}_3 = x_\zeta \mathbf{i} + y_\zeta \mathbf{j} + z_\zeta \mathbf{k}
\end{aligned} \tag{2.4.2}$$

It is important to note that the new basis vectors, $\mathbf{e}_i$, are in general neither normalized nor orthogonal. The location of the new basis vectors are depicted in Figure 2.4.3.

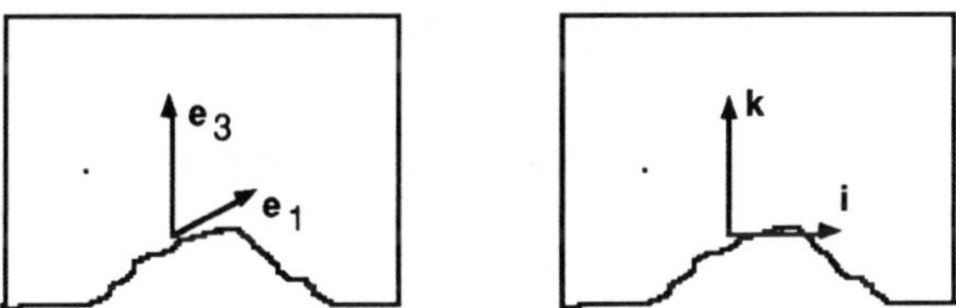

Figure 2.4.3: Basis vectors in the two coordinate systems

The $\mathbf{e}_i$ are needed to formulate boundary conditions in the transformed system.

The transformation

$$\begin{aligned}
x &= x(\xi, \eta, \zeta) & & & \xi &= \xi(x, y, z) \\
y &= y(\xi, \eta, \zeta) & &\longleftrightarrow & \eta &= \eta(x, y, z) \\
z &= z(\xi, \eta, \zeta) & & & \zeta &= \zeta(x, y, z)
\end{aligned} \tag{2.4.3}$$

shall be unique, i.e. the Jacobian

$$J = \frac{\partial(x,y,z)}{\partial(\xi,\eta,\zeta)} \tag{2.4.4}$$

shall be non-zero. The positive sign of the Jacobian guarantees the same orientation of the transformed coordinates. The type of equations cannot be changed by coordinate transformations.

The transformation matrix needed for numerical work is easily derived from eq. (2.4.3) (e.g. $\xi = \xi(x,y,z) \rightarrow d\xi = \xi_x dx + \xi_y dy + \xi_z dz$):

$$\begin{Bmatrix} d\xi \\ d\eta \\ d\zeta \end{Bmatrix} = \begin{pmatrix} \xi_x & \xi_y & \xi_z \\ \eta_x & \eta_y & \eta_z \\ \zeta_x & \zeta_y & \zeta_z \end{pmatrix} \begin{Bmatrix} dx \\ dy \\ dz \end{Bmatrix} \quad \longleftrightarrow \quad \begin{Bmatrix} dx \\ dy \\ dz \end{Bmatrix} = \begin{pmatrix} x_\xi & x_\eta & x_\zeta \\ y_\xi & y_\eta & y_\zeta \\ z_\xi & z_\eta & z_\zeta \end{pmatrix} \begin{Bmatrix} d\xi \\ d\eta \\ d\zeta \end{Bmatrix} \tag{2.4.5}$$

With $\mathbf{r} = \mathbf{r}(\xi,\eta,\zeta)$, ... given, one calculates from eq. (2.4.5) the transformation matrix:

$$\begin{pmatrix} \xi_x & \xi_y & \xi_z \\ \eta_x & \eta_y & \eta_z \\ \zeta_x & \zeta_y & \zeta_z \end{pmatrix} = \frac{1}{J} \begin{pmatrix} y_\eta z_\zeta - y_\zeta z_\eta & x_\zeta z_\eta - x_\eta z_\zeta & x_\eta y_\zeta - x_\zeta y_\eta \\ y_\zeta z_\xi - y_\xi z_\zeta & x_\xi z_\zeta - x_\zeta z_\xi & x_\zeta y_\xi - x_\xi y_\zeta \\ y_\xi z_\eta - y_\eta z_\xi & x_\eta z_\xi - x_\xi z_\eta & x_\xi y_\eta - x_\eta y_\xi \end{pmatrix} \tag{2.4.6}$$

where the Jacobian is given by

$$J = x_\xi y_\eta z_\zeta - x_\eta y_\xi z_\zeta - x_\xi y_\zeta z_\eta + x_\eta y_\zeta z_\xi + x_\zeta y_\xi z_\eta - x_\zeta y_\eta z_\xi \,. \tag{2.4.7}$$

Transformed Equations. The dynamical equations are easily transformed into a different coordinate system by starting from the flux-form of the quations. A vector equation is said to be in strong flux-form if it can be written as:

$$\frac{\partial}{\partial t}\mathbf{a} + \frac{\partial}{\partial x}\mathbf{b} + \frac{\partial}{\partial y}\mathbf{c} + \frac{\partial}{\partial z}\mathbf{d} = \mathbf{e}\,. \tag{2.4.8}$$

Replacing the Cartesian derivatives by their respective derivatives in the transformed system, e.g.

$$\frac{\partial}{\partial x} = \xi_x \frac{\partial}{\partial \xi} + \eta_x \frac{\partial}{\partial \eta} + \zeta_x \frac{\partial}{\partial \zeta} \tag{2.4.9}$$

one obtains from eq. (2.4.8)

$$\frac{\partial}{\partial t}\mathbf{a} + \{\xi_x \mathbf{b}_\xi + \xi_y \mathbf{c}_\xi + \xi_z \mathbf{d}_\xi\} + \{\eta_x \mathbf{b}_\eta + \eta_y \mathbf{c}_\eta + \eta_z \mathbf{d}_\eta\} + \{\zeta_x \mathbf{b}_\zeta + \zeta_y \mathbf{c}_\zeta + \zeta_z \mathbf{d}_\zeta\} = \mathbf{e}\,. \tag{2.4.10}$$

Equation (2.4.10) is multiplied by J and, after partial integration of the contributions of the form

$$J\xi_x \mathbf{b}_\xi = (J\xi_x \mathbf{b})_\xi - \mathbf{b}_\xi (J\xi_x)_\xi \,, \tag{2.4.11}$$

one obtains

$$\begin{aligned}\frac{\partial}{\partial t}(J\mathbf{a}) &+ \frac{\partial}{\partial \xi}\{J\xi_x\mathbf{b} + J\xi_y\mathbf{c} + J\xi_z\mathbf{d}\} + \frac{\partial}{\partial \eta}\{J\eta_x\mathbf{b} + J\eta_y\mathbf{c} + \eta_z\mathbf{d}\} \\ &+ \frac{\partial}{\partial \zeta}\{J\zeta_x\mathbf{b} + J\zeta_y\mathbf{c} + J\zeta_z\mathbf{d}\} - \mathbf{b}\{(J\xi_x)_\xi + (J\eta_x)_\eta + (J\zeta_x)_\zeta\} \\ &- \mathbf{c}\{(J\xi_y)_\xi + (J\eta_y)_\eta + (J\zeta_y)_\zeta\} - \mathbf{d}\{(J\xi_z)_\xi + (J\eta_z)_\eta + (J\zeta_z)_\zeta\} = J\mathbf{e}\,.\end{aligned} \tag{24.12}$$

The contributions with negative signs contain as factors the Jacobi identities which are identically zero. Therefore

$$\begin{aligned}\frac{\partial}{\partial t}(J\mathbf{a}) &+ \frac{\partial}{\partial \xi}\{J\xi_x\mathbf{b} + J\xi_y\mathbf{c} + J\xi_z\mathbf{d}\} + \frac{\partial}{\partial \eta}\{J\eta_x\mathbf{b} + J\eta_y\mathbf{c} + \eta_z\mathbf{d}\} \\ &+ \frac{\partial}{\partial \zeta}\{J\zeta_x\mathbf{b} + J\zeta_y\mathbf{c} + J\zeta_z\mathbf{d}\} = J\mathbf{e}\,.\end{aligned} \tag{2.4.13}$$

As an example the flux-form of eqs. (2.1.30) is transformed leaving out diffusion and Coriolis force. For the different vectors **a**, **b**, **c**, **d** and **e** one finds:

$$\mathbf{a} = \left\{\begin{array}{c}\overline{\rho}u \\ \overline{\rho}v \\ \overline{\rho}w \\ \overline{\rho}\theta \\ 0\end{array}\right\}; \quad \mathbf{b} = \left\{\begin{array}{c}\overline{\rho}uu + p' \\ \overline{\rho}uv \\ \overline{\rho}uw \\ \overline{\rho}uT \\ \overline{\rho}u\end{array}\right\}; \quad \mathbf{c} = \left\{\begin{array}{c}\overline{\rho}vu \\ \overline{\rho}vv + p' \\ \overline{\rho}vw \\ \overline{\rho}vT \\ \overline{\rho}v\end{array}\right\};$$

$$\mathbf{d} = \left\{\begin{array}{c}\overline{\rho}wu \\ \overline{\rho}wv \\ \overline{\rho}ww + p' \\ \overline{\rho}wT \\ \overline{\rho}w\end{array}\right\}; \quad \mathbf{e} = \left\{\begin{array}{c}0 \\ 0 \\ (g\overline{\rho}/\overline{\theta})\theta' \\ 0 \\ 0\end{array}\right\} \tag{2.4.14}$$

For the coordinate system it is assumed that the x- and y- coordinates remain unchanged, except for variable scaling, and only the z-coordinate is transformed to follow the topography. The transformation matrix (2.4.6) simplifies to

$$\begin{pmatrix}\xi_x & \xi_y & \xi_z \\ \eta_x & \eta_y & \eta_z \\ \zeta_x & \zeta_y & \zeta_z\end{pmatrix} = \begin{pmatrix}\xi_x & 0 & 0 \\ 0 & \eta_y & 0 \\ \zeta_x & \zeta_y & \zeta_z\end{pmatrix} \tag{2.4.15}$$

Eqs. (2.4.14) are inserted into eq. (2.4.13), and the transformed components can be read off again:

$$\begin{aligned}\frac{\partial}{\partial t}(J\overline{\rho}u) &+ \frac{\partial}{\partial \xi}\{J\xi_x(\overline{\rho}u^2 + p')\} + \frac{\partial}{\partial \eta}\{J\eta_y\overline{\rho}vu\} \\ &+ \frac{\partial}{\partial \zeta}\{J[\zeta_x(\overline{\rho}u^2 + p') + \zeta_y\overline{\rho}vu + \zeta_z\overline{\rho}wu]\} = 0\end{aligned}$$

$$\frac{\partial}{\partial t}(J\overline{\rho}v) + \frac{\partial}{\partial \xi}\{J\xi_x\overline{\rho}uv\} + \frac{\partial}{\partial \eta}\{J\eta_y(\overline{\rho}v^2 + p')\}$$
$$+ \frac{\partial}{\partial \zeta}\{J[\zeta_x\overline{\rho}uv + \zeta_y(\overline{\rho}v^2 + p') + \zeta_z\overline{\rho}wv]\} = 0$$

$$\frac{\partial}{\partial t}(J\overline{\rho}w) + \frac{\partial}{\partial \xi}\{J\xi_x\overline{\rho}uw\} + \frac{\partial}{\partial \eta}\{J\eta_y\overline{\rho}vw\}$$
$$+ \frac{\partial}{\partial \zeta}\{J[\zeta_x\overline{\rho}uw + \zeta_y\overline{\rho}vw + \zeta_z(\overline{\rho}w^2 + p')]\} = g\frac{\overline{\rho}}{\overline{\theta}}\theta'$$

$$\frac{\partial}{\partial t}(J\overline{\rho}\theta) + \frac{\partial}{\partial \xi}\{J\xi_x\overline{\rho}u\theta\} + \frac{\partial}{\partial \eta}\{J\eta_y\overline{\rho}v\theta\}$$
$$+ \frac{\partial}{\partial \zeta}\{J[\zeta_x(\overline{\rho}u\theta) + \zeta_y(\overline{\rho}v\theta + \zeta_z(\overline{\rho}v\theta]\} = 0$$

$$\frac{\partial}{\partial \xi}\{J\xi_x\overline{\rho}u\} + \frac{\partial}{\partial \eta}\{\eta_y\overline{\rho}v\} + \frac{\partial}{\partial \zeta}\{J[\zeta_x\overline{\rho}u + \zeta_y\overline{\rho}v + \zeta_z\overline{\rho}w]\} = 0 \qquad (2.4.16)$$

Introducing the abbreviations (which are in fact the contravariant velocities)

$$\mathbf{U} = \begin{Bmatrix} U \\ V \\ W \end{Bmatrix} = \begin{Bmatrix} \xi_x u \\ \eta_y v \\ \zeta_x u + \eta_y v + \zeta_z w \end{Bmatrix}, \qquad \nabla_\xi = \begin{Bmatrix} \partial/\partial\xi \\ \partial/\partial\eta \\ \partial/\partial\zeta \end{Bmatrix} \qquad (2.4.17)$$

the equations have the simple form:

$$\begin{aligned}
\frac{\partial}{\partial t}(J\overline{\rho}u) + \nabla_\xi \cdot (\mathbf{U}J\overline{\rho}u) + \{\frac{\partial}{\partial \xi}\xi_x + \frac{\partial}{\partial \zeta}\zeta_x\}Jp' &= 0 \\
\frac{\partial}{\partial t}(J\overline{\rho}v) + \nabla_\xi \cdot (\mathbf{U}J\overline{\rho}v) + \{\frac{\partial}{\partial \eta}\eta_y + \frac{\partial}{\partial \zeta}\zeta_y\}Jp' &= 0 \\
\frac{\partial}{\partial t}(J\overline{\rho}w) + \nabla_\xi \cdot (\mathbf{U}J\overline{\rho}w) + \frac{\partial}{\partial \zeta}\{J[\zeta_z p']\} &= g\frac{\overline{\rho}}{\overline{\theta}}\theta' \\
\frac{\partial}{\partial t}(J\overline{\rho}\theta) + \nabla_\xi \cdot (\mathbf{U}J\overline{\rho}\theta) &= 0 \\
\nabla_\xi \cdot (J\overline{\rho}\mathbf{U}) &= 0
\end{aligned} \qquad (2.4.18)$$

The flux-form of the equations (2.4.18) can only be achieved through the mixed representation: the transported quantities are Cartesian velocities, and the transporting momenta are contravariant quantities.

The boundary treatment in the transformed system is illustrated with the surface boundary condition for pressure:

$$\frac{\partial}{\partial \mathbf{n}}p'(\zeta = 0) = \mathbf{n} \cdot \nabla_\xi p'(\zeta = 0) = 0\,. \qquad (2.4.19)$$

$\mathbf{n}$ is the normal vector of the surface, and the Nabla Operator is given in equation (2.4.17):

$$\mathbf{n} \sim \mathbf{e}_1 \times \mathbf{e}_2 = J\left\{\xi_x\zeta_x\mathbf{e}_1 + \eta_y\zeta_y\mathbf{e}_2 + [(\zeta_x)^2 + (\zeta_y)^2 + (\zeta_z)^2]\mathbf{e}_3\right\}\,. \qquad (2.4.20)$$

Therefore

$$\left\{\xi_x\zeta_x\frac{\partial}{\partial\xi}+\eta_y\zeta_y\frac{\partial}{\partial\eta}+[(\zeta_x)^2+(\zeta_y)^2+(\zeta_z)^2]\frac{\partial}{\partial\zeta}\right\}p'(\zeta=0)=0\,. \tag{2.4.21}$$

In a similar way other boundary conditions are treated.

References

BURRIDGE, D. M. 1975 A split semi-implicit reformulation of the Bushby-Timpson 10-level model. *Quart. J. Roy. Meteorol. Soc.* **101**, 777-792

CULLEN, M.J.P. 1990 A test of a semi-implicit integration technique for a fully compressible non-hydrostatic model. *Quart. J. Roy. Meteorol. Soc.* **116**, 1253-1258

DAVIES, H. C. 1983 Limitations of some common lateral boundary schemes used in regional NWP models. *Mon. Wea. Rev.* **111**, 1002-1011

DEARDORFF, J. W. 1972 Parameterization of the planetary boundary layer for use in general circulation models. *Mon. Wea. Rev.* **100**, 93-106

DURRAN, D. R. & YANG, M.-Y. 1993 Toward more accurate wave-permeable boundary conditions. *Mon. Wea. Rev.* **121**, 604-620

DURRAN, D. R. 1989 Improving the anelastic approximation. *J. Atmos. Sci.* **46**, 1453-1461

DUTTON, J.A. & FICHTL, G. H. 1969 Approximate equations of motion for gases and liquids. *J. Atmos. Sci.* **26**, 241-254

ENGQUIST, B. & MAJDA, A. 1977 Absorbing boundary conditions for the numerical simulation of waves. *Mathematics of Computation* **31**, 629-651

FOREMAN, M. G. G. 1986 An Accuracy analysis of boundary conditions for the forced shallow Water equations. *J. Comp. Phys.* **64**, 334-367

GIVOLI, D. 1991 Non-reflecting boundary conditions. *J. Comp. Phys.* **94**, 1-29

GRESHO, P. M. 1991 Incompressible fluid dynamics: some fundamental formulation issues. *Ann. Rev. Fluid. Mech.* **23**, 413-453

GRESHO, P. M. 1987 On pressure boundary conditions for the incompressible Navier-Stokes equations. *Int. J. Num. Meth. Fluids* **7**, 1111-1145

GYARMATI, 1970 Non-equilibrium thermodynamics. *Springer Verlag*, Berlin

HADAMARD, J. 1921 Lectures on Cauchy's Problem in Linear Partial Differential Equations. *Yale University*, Reprint Dover, New York 1956

HEDSTROM, G. W. 1979 Nonreflecting boundary conditions for nonlinear hyperbolic systems. *J. Comp. Phys.* **30**, 222-237

KLEMP, J. P. & DURRAN, D. R. 1983 An upper boundary condition permitting internal gravity wave radiation in numerical mesoscale models. *Mon. Wea. Rev.* **111**, 430-444

KLAINERMAN, S. & MAJDA, A. 1981 Singular limits of quasilinear hyperbolic systems with large parameters and the incompressible limit of compressible fluids. *Com. Pure Appl. Math.* **XXXIV**, 481-524

KLAINERMAN, S. & MAJDA, A. 1982 Compressible and incompressible fluids. *Com. Pure Appl. Math.* **XXXV**, 629-651

KREISS, H.-O. & LORENZ, A. 1989 Initial-boundary value problems and the Navier-Stokes equations. *Academic Press*, Boston

KRÖNER, D. 1991 Absorbing boundary conditions for the linearized Euler Equations in 2-d. *Mathematics of Computing,* **57**, 153-167

NIRENBERG, L. 1973 Lectures on linear partial differential equations. *C.B.M.S. Regional Conf. Ser. in Math.* **no 17**, Providence, Rhode Island

OGURA, Y. & PHILLIPS, N.A. 1962 Scale analysis of deep and shallow convection in the atmosphere. *J. Atmos. Sci.* **19**, 173-179

ORLANSKI, I. 1976 A simple boundary condition for multi-dimensional flows. *J. Comp. Phys.* **21**, 251-269

RAYMOND, W. H. & KUO, H. L. 1984 A radiation boundary condition for multi-dimensional flow. *Quart. J. Roy. Meteorol. Soc.* **110**, 535-551

TATSUMI, Y. 1980 Comparison of the time-dependent lateral boundary conditions proposed by Davies and Hovermale. *WGNE Progress Rep. No. 21*, WMO Secretariat 93-94

WIPPERMANN, F. 1981 The applicability of several approximations in meso-scale modelling - a linear approach. *Contrib. Phys. Atmos.* **54**, 298-308

Address of the authors

Dr. Dieter P. Eppel
Dr. Ulrich Callies
Atmospheric Physics Division
GKSS-Research Centre
Max-Planck-Str. 1
D-21 502 Geesthacht

III Thermodynamic and Radiative Processes in the Atmosphere

Martin Beniston[1] and Johannes Schmetz[2]

3.1 Introduction

Numerous factors need to be taken into account when investigating regional scale meteorological processes; these are essentially linked to the dynamic and thermal characteristics of the atmosphere. Air flow will determine the speed and direction with which a meteorological variable will be transported and dispersed, while the thermal structure of the atmosphere will control the nature of the dispersion through local stability or instability, such as the presence of inversion layers which act as a "lid" and strongly inhibit vertical motion. Inversion situations are especially important for air quality problems, as poor dispersion and trapping of pollutants beneath the inversion will inevitably lead to poor local and regional air quality. Atmospheric dynamics and thermodynamics are influenced by various factors, where complex feedbacks and interactions occur between the fundamental meteorological processes and other elements of the terrestrial system; among these elements, the following are perhaps the most important:

a. The nature of the underlying surface. The presence of mountain and valley systems exerts a particularly strong influence on air flow, through channeling effects of the orography, and through the generation of mountain and valley breezes according to local thermal criteria. Land use characteristics are also a major factor in atmospheric flow dynamics, as surface-induced friction effects are a function of the type of land cover (urban areas, forests, water surfaces, grasslands, etc.). The thermodynamic structure of the Atmospheric Boundary Layer (ABL) is also a function of surface temperature and moisture heterogeneities, related to vegetation type and the presence of rivers, lakes or oceans.

b. Atmospheric turbulence. Turbulence encompasses random motion within a given air flow. The intensity of turbulence can depend on purely dynamic factors, in particular surface roughness or vertical flow deformation (wind shear), or thermal factors such as atmospheric stability, or a combination of both factors. Strong turbulence in the ABL can significantly perturb flow, temperature, and moisture characteristics of the regional atmosphere, with significant implications for air quality and short-term forcing on the climate system.

c. Solar, terrestrial, and atmospheric radiation. Solar radiation intercepted by the earth is the principal external source of energy which drives the atmospheric and climate machine. Globally averaged over one year, about 44% of the solar energy is absorbed by the earth's surface, 26% is absorbed by the atmosphere, and 30% is reflected to space. The solar radiative energy absorbed at the surface is compensated for by longwave emission and fluxes of latent and sensible heat, thereby warming the lower layers of the atmosphere. Absorbing gases, especially water vapor and carbon dioxide as well as clouds, intercept the upwelling long-wave radiation and re-emit at their ambient temperature. Since re-emission typically takes place at lower temperatures, absorbing gase and clouds tend to trap the longwave loss of the earth's surface, an effect commonly referred to as the "Greenhouse Effect" (although the analogy is not quite correct, since a greenhouse inhibits the loss of energy by convective processes).

d. Cloud activity. Cloud formations, especially of the turbulent cumulus type, are capable of producing significant dynamic and thermodynamic modifications to the regional atmosphere. The formation of cloud condensation droplets is accompanied by a release of latent heat; at the cloud edges and cloud top, droplet evaporation leads to latent heat absorption and corresponding cooling of the neighboring cloud-free air. Dynamically, a

A. Gyr and F-S. Rys (eds.), Diffusion and Transport of Pollutants in Atmospheric Mesoscale Flow Fields, 57–88.

cumulus-type cloud is an unstable and turbulent phenomenon; exchange of air at the cloud boundaries can result in secondary circulations which can perturb atmospheric flows from the surface through to the upper troposphere. The interaction between clouds and radiation in climate models is a rather crudely parameterized feature. Further observational studies, varying from the global scale satellite programs to small-scale aircraft studies are required to better understand the role of clouds in weather and climate; in addition, more detailed cloud/radiation interaction studies with mesoscale atmospheric models are needed to capture the underlying mechanisms on the regional scale and which could ultimately provide improvements to predictions of global change.

e. Precipitation processes. The triggering of precipitation is of course closely linked to cloud formation. In a cumulus or a cumulo-nimbus cloud, precipitation is the sign of a stabilization or decay of cloud growth, and therefore an attenuation of the cloud dynamic and thermodynamic influence on its immediate environment. As rainwater exits its saturated cloud environment, it begins to evaporate and therefore cools the sub-cloud air layers. In the case of moderate to heavy rain, the combined effects of evaporational cooling and rainwater fallout lead to a reversal of vertical motion, cutting the cloud off from its low-level dynamic and moisture sources and leading to its eventual decay. These features of precipitation can significantly modify ABL processes.

A numerical model of regional atmospheric processes should take into account as many of these processes as possible in order to provide coherent simulated data. This chapter will therefore provide an overview of some of the theory underlying atmospheric thermodynamics and radiation, and will provide some examples of the application of these physical principles to phenomena which range from the mesoscale to the macroscale range of meteorological processes. Section 3 provides a brief introduction to atmospheric radiative transfer, and Section 4 deals with the effect of clouds on the radiation budget; there we will somewhat depart from our general tendency to emphasize issues that are important to mesoscale meteorology. Section 5 draws on previous work where the effects of clouds on radiation were investigated with the use of a mesoscale atmospheric model.

3.2. Atmospheric thermodynamics for dry, moist, and saturated conditions

The objective of the first part of this chapter is to introduce the reader to the applications of the principal laws of thermodynamics to the atmosphere. This section will therefore briefly define the first two laws of thermodynamics and the equation of state, and then go on to apply the principles contained therein to atmospheric processes in both a dry atmosphere and one in which clouds are present. The section will end with the formulation of the conservation equations for temperature and moisture, which are the essential relations which may be modeled using numerical techniques. It will be seen that, in the conservation equation for temperature, an additional term representing radiative flux exchange is necesary to fully describe the adiabatic and non-adiabatic contributions to temperature tendencies. The equations thus defined will then feed into the section on atmospheric radiative transfer.

3.2.1 Principal laws and relationships

3.2.1.1 The Equation of State. The equation of state represents the simplest expression of a thermodynamic system in the absence of water vapor in the atmosphere, one in which gaseous composition is uniform and constant, and no chemical reactions between gaseous elements are taking place. In such a system, once the mass of the gaseous element has been specified, only two other independent variables are required to define the physical state of the gas, namely its pressure and temperature. The system is therefore described by the following relationship:

$$p\,\alpha = R\,T \;(\text{or } p = \rho\,R\,T) \tag{3.1}$$

with the standard symbols p representing pressure, a the specific volume, T the temperature, R the gas constant for a standard atmosphere (287 J/kg/K), and r the density. It is seen that a is the inverse of r.

3.2.1.2 The First Law of Thermodynamics. This law links the temperature of a gas to the kinetic energy of the molecules which make up the gas. The internal energy of a gas is defined as the molecular energy within a given volume of gas. The internal energy can be modified by changing the temperature of the gas, or by making the gas work against its environment (through expansion or contraction of the gaseous volume), or by a combination of both processes. This can be described in mathematical terms through:

$$dE = dQ + dW \tag{3.2}$$

where dE is the change in internal energy of the gas, dQ is the change of heat, and dW is the change in work done by the gas in expanding or contracting; in general, work done by a gas is expressed in terms of pressure. Thermodynamic changes to a gaseous state can intervene either through processes in which the volume remains constant (i.e. through temperature changes only), or where the volume may change but where pressure remains constant. In the former case, the First Law may be written as:

$$dQ = M\, c_v\, dT \tag{3.3}$$

and in the latter case as:

$$dE = M\, c_v\, dT = M\, c_p dT - p\, dV \tag{3.4}$$

where c_v is the specific heat of the gas at constant volume, c_p is the specific heat at constant volume, V is the volume. Equation 3.4 may be re-written as:

$$p\ dV = M\, dT\, (c_v - c_p) \tag{3.5}$$

Since $R = c_v - c_p$, simple manipulation of Equation 3.5 reads:

$$p\, dV = M\, R\, dT \tag{3.6}$$

The reader is refered to standard introductory textbooks on atmospheric physics, such as Holton (1972) or Pielke (1984) for a more complete description of the equations defined here.

3.2.1.3 Potential Temperature. Manipulation of Equations 3.5 and 3.6 leads to two forms of the heat change equation dQ for processes at constant volume or at constant pressure:

$$dQ = c_v\, dT + p\, d\alpha \tag{3.7}$$

$$dQ = cpdT - \alpha\, dp \tag{3.8}$$

Dividing the above equation by temperature T, one obtains:

$$\frac{dQ}{T} = c_p \frac{dT}{T} - \alpha \frac{dp}{T} \tag{3.9}$$

which, when substituting the equation of state in the second term of the right-hand-side of Equation 3.9, reads:

$$\frac{dQ}{T} = c_p \frac{dT}{T} - R \frac{dp}{p} = c_p\, d \ln (T\, p^{-R/cp}) \tag{3.10}$$

If dQ = 0, the quantity $T\,p^{-R/cp}$ remains constant.The variable is said in this case to be conserved. Thermodynamic modifications to a gas where changes in heat are nil are known as adiabatic processes i.e., there is no addition or extraction of heat to the system. A new variable can be defined as a result of adiabatic atmospheric processes: potential temperature. Constancy of $T\,p^{-R/cp}$ implies that:

$$T\,p^{-R/cp} = T_0\,p_0^{-R/cp} \tag{3.11}$$

where T_0 and p_0 are reference values of temperature and pressure, respectively. If the reference temperature is defined as potential temperature θ, the formal definition of this temperature reads:

$$\theta = T\left(\frac{p_0}{p}\right)^{R/cp} \tag{3.12}$$

Here, the ratio R/c_p is approximately constant at 0.286 for most atmospheric processes, and the reference pressure is taken at 10^5 Pa (1000 mb), which is a near ground-level value of pressure. Potential temperature is the temperature which a parcel of air at any given level would attain if it were brought adiabatically to the 1000 mb level. The importance of this variable in atmospheric dynamics and thermodynamics will be described in a later section of this text.

3.2.1.4 The Second Law of Thermodynamics. It follows from Equation 3.10 that:

$$dQ = c_p T\, d(\ln\theta) \tag{3.13}$$

where the definition of potential temperature has replaced $T\,p^{-R/cp}$. If a gas changes its state from an initial phase I to a final phase 2, the heat added to the system Q_{1-2} is given by the integral of Equation 3.13, i.e.:

$$Q_{1-2} = c_p \int T\, d(\ln\theta) = c_p \int T \frac{d\theta}{\theta} \tag{3.14}$$

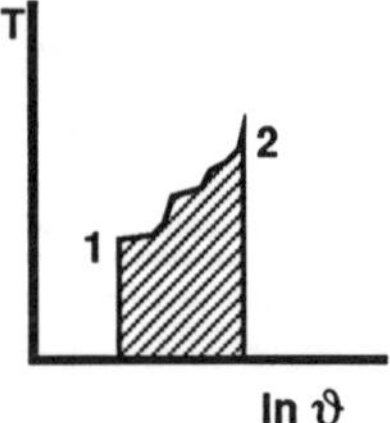

Figure 3.1: Change in thermodynamic state of a gas

Figure 3.1 illustrates the change in thermodynamic state of a gas described by Equations 3.13 and 3.14. The change in state is given by the line linking 1-2, while the heat added to the system is the area beneath the curve. It is seen from Figure 3.1 that in order to change the state of a gas, it is not sufficient to know only its initial and final states, but also the different states it encounters as it changes from its initial to its final state, i.e., curve 1-2. It is obvious that an infinite number of curves may link the two states. Consequently the heat added to or removed from a volume of gas is not a function of its state, but of the change in state of the gas. If a quantity S is defined, such that:

$$dS = c_p\, d\ln\theta \quad (\text{or } dS = \frac{dQ}{T}) \tag{3.15}$$

the integral of the above equation reads:

$$S = S_0 + c_p \ln \theta \tag{3.16}$$

where S_0 is an arbitrary constant. S is known as the *entropy* of the system.

Where the first law of thermodynamics stated that an energy relation exists for each process considered, the Second Law goes further by identifying whether, and in what manner, a thermodynamic process may take place.

3.2.2 Applications of the laws of thermodynamics to the atmosphere

It will be seen that certain fundamental relations, useful in atmospheric physics, can be obtained by manipulations and combinations of the various thermodynamic principles which have been covered in this chapter.

3.2.2.1 The Adiabatic Temperature Gradient. The hydrostatic equation can be altered through applications of the equation of state and the first law of thermodynamics to define the characteristic vertical temperature profile in the atmosphere. The hydrostatic equation relates the vertical pressure gradient and density through the gravitational acceleration term in the following manner:

$$\frac{dp}{dz} = -\rho g \tag{3.17}$$

Using the equation of state to substitute for density, Equation 3.17 reads:

$$\frac{dp}{dz} = -\frac{pg}{RT} \tag{3.18}$$

The first law of thermodynamics in the form given by Equation 3.8 reduces to:

$$c_p \, dT = \alpha dp \tag{3.19}$$

for an adiabatic process where dQ=0. dp in the above equation may be replaced by the formulation in Equation 3.18, such that:

$$\rho \, c_p \, dT = -\frac{pg}{RT} \, dz \tag{3.20}$$

which, after simple manipulation, reads:

$$\frac{dT}{dz} = -\frac{pg}{\rho c_p RT} \tag{3.21}$$

Replacing once again density in Equation 3.21 by its formulation from the equation of state, the terms ρ, p, R, and T cancel out so that the above equation is simplified to:

$$\frac{dT}{dz} = -\frac{g}{c_p} \tag{3.22}$$

The above relationship describes the vertical temperature gradient for an adiabatic, hydrostatic, and dry atmosphere, and is known as the *dry adiabatic lapse rate*. For most atmospheric applications, g and c_p are constants (9.807m / s^2 and 1005 J / kg / K, respectively) so that the vertical temperature gradient is approximately - 9.8K / km or a decrease of nearly 1° C for every 100 m rise in the atmosphere. It will be seen that departures from this gradient define conditions of atmospheric stability or instability.

3.2.3 Thermodynamic effects of atmospheric humidity

So far only thermodynamic effects representative of a dry atmosphere have been considered. What makes the atmosphere perhaps the most unique of fluids is the fact that water vapor exists within the air, and that this moisture can undergo phase changes; by doing so, the thermodynamic structure of the atmosphere may be profoundly modified. Phase changes of water include processes such as evaporation (conversion of liquid water to vapor); sublimation (transformation of ice to vapor); and melting (conversion from the ice phase to the liquid phase). The reverse processes, vapor liquid (condensation), liquid-ice (freezing), and vapor-ice (drystalization), all occur naturally in the atmosphere.

Figure 3.2 illustrates schematically the thermodynamic changes which can occur in the vicinity of clouds when condensation, crystalization, and evaporation processes are taking place. During condensation, latent heat is released at the rate of 2.45 x 10^6 J for each kg of moist air condensed; this heat is absorbed at the same rate for evaporation processes, but what is important to note here is that in doing so, the cloud transforms thermodynamically a vertical segment of the atmosphere whose depth is proportional to the vertical dimensions of the cloud; the process is in this case irreversible, i.e., the initial state will not be found again once condensation and evaporation have taken place successively.

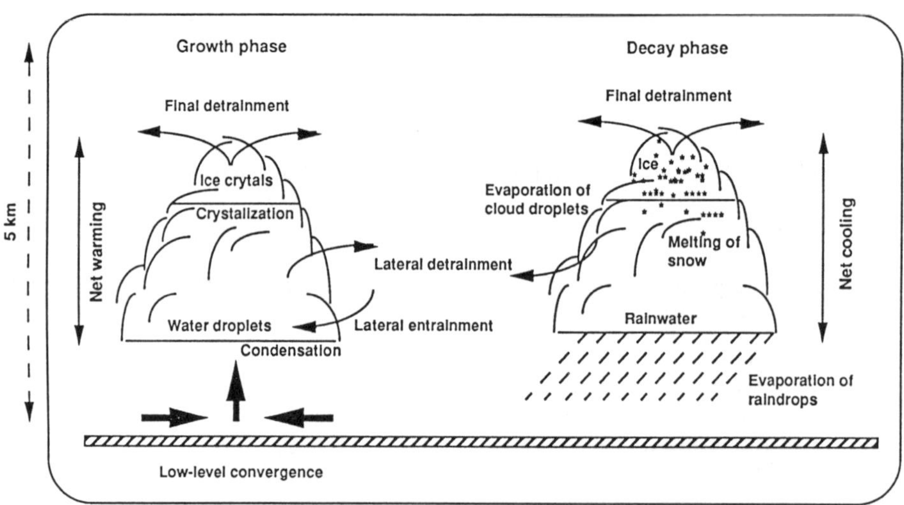

Figure 3.2: Illustration of thermodynamic mechanisms associated with cloud activity

Atmospheric water represents only between 0 and 3 % of the total volume of the atmosphere, but is one of the most remarkable components in shaping day-to-day weather. Among the major manifestations of the presence of water in the atmosphere, one can mention clouds, fog and mist, dew, frost, ice, and precipitation in its many forms (rain, drizzle, snow, hail, etc.) All these features play an important role in economy of a country in particular agriculture and forestery, transportation, acid pollution; etc.

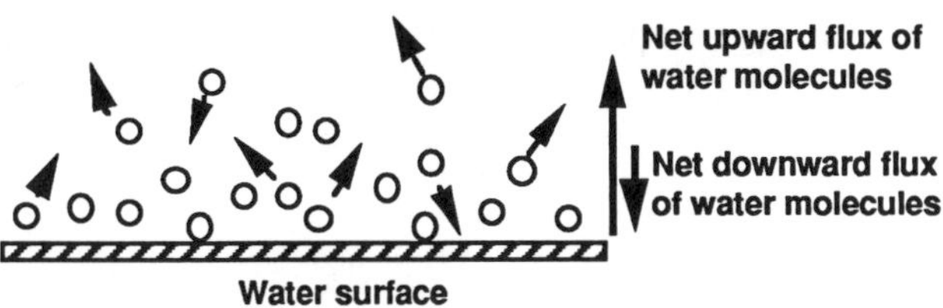

Figure 3.3: Evaporation of molecules at a water-air interface

3.2.3.1 Partial water vapor pressure. The sketch in Figure 3.3 illustrates molecular activity at the interface between the air and a body of water in the process of evaporation. Since the number of molecules leaving the water surface exceed those arriving towards the water surface, there exists a tension due to water-vapor; this tension is measured in the form of pressure. Indeed, a sensitive pressure-gauge in the vicinity of an evaporating water surface would record a slight increase in atmospheric pressure which is due to the contribution of water vapor molecules entering the atmosphere. This pressure is known as the *partial water vapor pressure*, or simply *partial pressure*: The partial pressure will increase until it is equal to the water vapor tension at the air-water interface. When this equilibrium state is reached, no further evaporation is possible; furthermore, if the air and water are at the same temperature, the air at this equilibrium state is said to be *saturated.* Saturation vapor pressure is defined as the partial pressure of vapor when a state of neutral equilibrium is reached between a pure water surface and air at the same temperature. This pressure varies with temperature, so that for a higher air and water temperature, a greater vapor pressure is required in order to reach saturation. It is possible to define an equation of state for an atmosphere in the presence of water vapor, since this quantity is also measured in terms of pressure in the same manner as atmospheric pressure. If e represents partial water vapor pressure, then the quantity p - e represents dry atmospheric pressure (p is taken here as the total pressure including partial pressure effects). The equation of state for water vapor takes an analogous form to its dry atmospheric counterpart, i.e.:

$$e = \rho_v R_v T \tag{3.23}$$

where R_v and ρ_v are the water vapor gas constant and density, respectively. For dry air, the equation reads:

$$p - e = \rho_d R_d T \tag{3.24}$$

In these equations, the gas constants are defined, using the universal gas constant R^*, as:

$$R_v = \frac{R^*}{m_v}; \; R_{vd} = \frac{R^*}{m_d} \tag{3.25}$$

where $m_d = 28.8$ and $m_v = 17.9$. The ratio of the molecular weights of water vapor and dry air is therefore:

$$\varepsilon = \frac{m_v}{m_d} \cong 0.622 \tag{3.26}$$

2.3.2 Specific humidity. Specific humidity is defined as the mass of moist air contained within a unit mass of air (dry air + water vapor), such that:

$$q = \frac{\rho_v}{\rho_v + \rho_d} \tag{3.27}$$

Substituting for ρ_v and ρ_d from the equations of state 3.23 and 3.24, it is seen that:

$$q = \frac{m_v\, e}{m_d p_d + m_v e} \tag{3.28}$$

Dividing throughout by m_d yields:

$$q = \frac{\frac{m_v}{m_d}\, e}{p_d + \frac{m_v}{m_d}\, e} \tag{3.29}$$

Replacing p_d by $p - e$ and $\frac{m_v}{m_d}$ by ε, Equation 3.29 becomes:

$$q = \frac{\varepsilon\ e}{p - (1 - \varepsilon)\ e} \cong \frac{\varepsilon\, e}{p} \tag{3.30}$$

For most atmospheric applications, the last term of the equation above is a reasonable approximation, since $p >> e$. When the partial pressure reaches its saturation value e_s, the saturation specific humidity q_s is defined as:

$$q_s = \frac{\varepsilon\ e_s}{p - (1 - \varepsilon)\ e_s} \cong \frac{\varepsilon\ e_s}{p} \tag{3.31}$$

This leads to the definition of a further quantity frequently used in meterology, the relative humidity U:

$$U = 100\,\frac{e}{e_s} = 100\,\frac{q}{q_s}\ \text{, expressed in percent} \tag{3.32}$$

Relative humidity is in fact the ratio of observed partial vapor pressure to saturation vapor pressure, and not, as is often thought, the percentage of humidity in a given volume.

If a new thermodynamic quantity is defined to take into account the presence of moisture in the atmosphere, the virtual temperature T_v reads:

$$T_v = \frac{T\,(1 + \frac{q}{\varepsilon})}{1 + q} \cong T\,(1 + 0.61\ q) \tag{3.33}$$

so that the equation of state for a moist atmosphere is now:

$$p = \rho\, R\, T_v \tag{3.34}$$

It is clear that if there is no humidity present in a parcel of air, T_v will be identical to T and Equation 3.34 will reduce to the formulation for a dry atmosphere given by Equation 3.1.

3.2.4 Phase changes of water

Water is one of the only substances in the atmosphere which can exist in three phases - vapor, liquid, and solid - separately or even simultaneously. During a transition from one phase to another, large quantities of heat can be released or absorbed. The limits of phase changes are given by the equilibrium levels defined by the saturation vapor pressure, e_s, which is temperature-dependent as previously mentionned. The graph in Figure 3.4 illustrates the dependency between e_s and T for typical tropospheric values of temperature.

The reason for the dashed curve to the left of the ordinate is that pure liquid water can exist down to temperature as low as -39°C. A liquid water droplet will not freeze spontaneously unless nucleating agents are present. To the right of the solid line, a parcel of air would always be unsaturated with respect to water; however, in the small zone between the solid and dashed lines, a parcel of air would be unsaturated with respect to water but supersaturated with respect to ice. In this case, ice crystals would tend to form at the expense of any water droplets which may be present in this region. The formal relation between saturation vapor pressure and temperature, not to be derived here but which is based on entropy considerations, is given as:

$$e_s = e_{s0} \exp\left[\frac{\varepsilon L_v}{R}\left(\frac{1}{T_0} - \frac{1}{T}\right)\right] \tag{3.35}$$

where e_{s0} is the "triple-point" (611 pa) and T_0 is the temperature at the triple point (273.16°K).

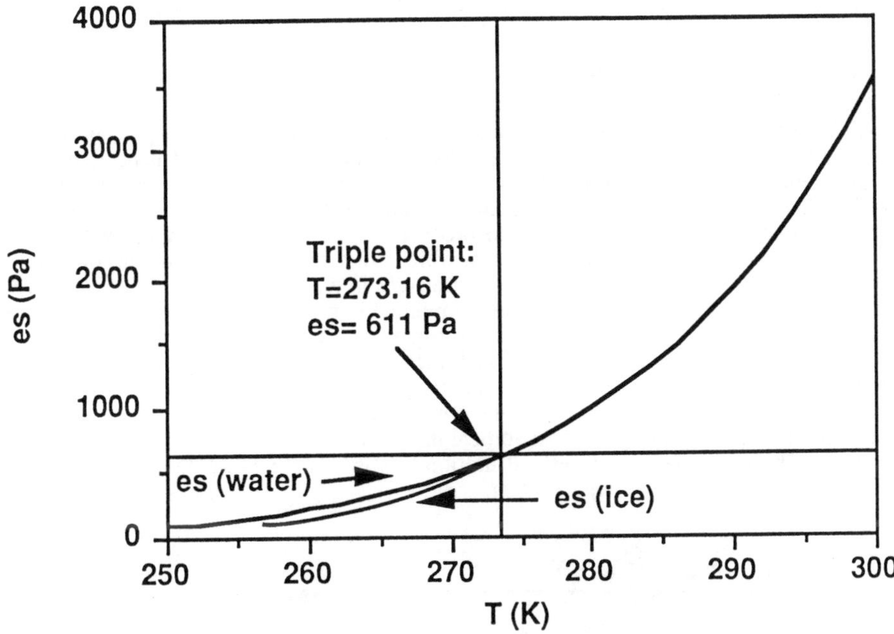

Figure 3.4: Relationship between saturation vapor pressure and temperature

3.2.4.1 The moist (saturated) adiabatic lapse rate. Cooling of rising air in a moist atmosphere is offset to a certain rate by latent heat released during condensation of water vapor. During ascent, if condensation occurs, the saturation specific humidity will change by a quantity dq_s. The heat released at this time is given by:

$$dQ = -L\, dq_s \tag{3.36}$$

where dq_s is a negative value due to condensation. Using Equation 3.8 and substituting for dQ, the equation for the first law of thermodynamics reads:

$$c_p\, dT - \alpha\, dp = -L\, dq_s \tag{3.37}$$

The chain rule of calculus then allows this last equation to be written as:

$$\frac{dT}{dz} = \frac{\alpha}{c_p}\frac{dp}{dz} - \frac{L}{c_p}\frac{dq_s}{dz} \tag{3.38}$$

By analogy with the derivation of dry adiabatic lapse rate, use of the hydrostatic equation and the equation of state yields the temperature gradient of air in which saturation is occuring:

$$\frac{dT}{dz} = -[\frac{g}{c_p} + \frac{L}{c_p}\frac{dq_s}{dz}] \tag{3.39}$$

The saturated adiabatic lapse rate is reduced with respect do the dry adiabatic lapse rate by a factor which is directly related to the amount of water vapor condensed into the rising air over the height interval Δz. Since q_s is termperature-dependent, it is clear that at high temperatures, where the atmosphere can hold considerable amounts of moisture, the saturation lapse rate can be as small as one-third of the dry lapse rate, while for cold atmospheric temperatures, there is little difference between the two lapse rates.

3.2.5 Atmospheric stability

Applying principles of fluid dynamics and thermodynamics to meteorology brings out the fact that vertical motion is one of the most essential parameters which influences short and medium term weather and climate. The importance of these vertical movements is linked to the fact that large quantities of heat and moisture are transported upwards or downwards in the atmosphere. These motions lead to processes such as condensation, evaporation, precipitation and cloud-associated features, which are major sinks and sources of heat. In comparaison to changes in temperature or moisture in horizontal flows, those associated with vertical motion are generally far more important in meteorological processes. Vertical accelerations are typically small compared to gravity and vertical pressure distribution; while this is true on the synoptic or regional (meso) scales for dry atmospheric conditions, the situation is different when thermodynamic processes are present. These can lead to vertical accelerations whose nature is function of the thermal gradient of the atmosphere. When thermodynamically-generated vertical motions are present, atmospheric conditions are said to be *unstable*; when vertical motion is absent or quickly damped by thermal effects, the atmosphere is *stably stratified.*

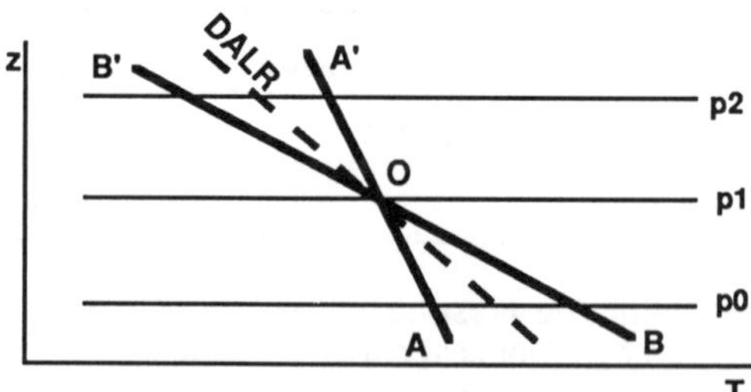

Figure 3.5: Lapse rates and atmospheric stability

Consider Figure 3.5, which represents three different vertical temperature profiles; the dotted line represents the dry adiabatic lapse rate (-9.8°K / km). Taking now the profile A - A' which is different from the adiabatic gradient for any number of reasons (cooling due to ground effects; warming above by condensation processes, for example), it may be seen that if a parcel of air is displaced upwards from point O, it will cool at the dry adiabatic rate in the absence of condensation; at the same level, the temperature of the parcel will be lower than that of the environment given by profile A - A'. Since the parcel is at the same pressure as its environment, considerations of the equation of state imply that the density of the parcel will be greater than that of its environment. In the absence of any other forces, the heavy air parcel will descend back to its original level O.

If, on the contrary, the parcel is displaced downwards from O, it will undergo a temperature rise at the dry adiabatic rate, leading to la less dense, therefore lighter, parcel. It will therefore tend to rise back towards its point of origin. Such a situation, represented by profile A - A' is an example of stable atmospheric conditions. Instability is represented by the vertical temperature profile B - B'. Repeating the exercise of displacing upwards a parcel of air, it is seen that on rising, its temperature will continuosly be greater than that of its environment, and hence the parcel's density will be lower. The parcel will therefore accelerate upward indefinitely until new thermal conditions are encountered. The reverse situation is also true for a parcel displaced downwards from point O: its density will be progressively greater than that of its environment, so that it will accelerate downward until it ultimately reaches a different atmospheric stratification near the ground surface. Unstable conditions are those which are most responsible for cloud formation and strong turbulence.

A third stability condition is given by the dry adiabatic lapse rate itself. If the environmental temperature profile were perfectly adiabatic, vertical motion in the absence of other forces would be neither enhanced nor inhibited. Such conditions are said to be *neutral*. In terms of potential temperature,a neutral temperature gradient is given by a zero vertical θ gradient, as seen in section 3.5.2; stable conditions are represented by a positive θ gradient, and unstable conditions are represented by a negative θ gradient. Mathematically, this is described by:

$$\frac{\partial \theta}{\partial z} > 0 \qquad \text{Stable conditions}$$

$$\frac{\partial \theta}{\partial z} = 0 \qquad \text{Neutral conditions}$$

$$\frac{\partial \theta}{\partial z} < 0 \qquad \text{Unstable conditions} \tag{3.40}$$

3.2.5.1 Stability conditions for a moist atmosphere. If a moist but unsaturated air parcel rises in the atmosphere, it will follow the dry adiabatic lapse rate until it reaches its condensation level. If the parcel continues to rise, it will follow the saturated lapse rate.

Figure 3.6 illustrates an example of a form of instability due to condensation effects; the observed environmental temperature profile is given by the dashed line, the dry adiabatic lapse rate by the solid line, and the saturation lapse rate by the solid curve. On ascending from its original level at O, a parcel of air rises till its condensation point A, at which point latent heat is released. At this juncture, the parcel rises along the saturated lapse rate until it interacts the environmental gradient at point B. Above this point, the parcel becomes perfectly unstable, since it is always warmer (less dense) than its environment. In order to rise from point O to point B through point A, ascent would need to be forced in some manner (for example by convergence of air at low levels leading to forced upward motion) until it reaches its point of transition between stable and unstable flow (point B), where vertical motion becomes spontaneous. Such a situation is known as *conditional instability* because a first condition of forced upward motion is required in order to reach free instability. Such conditions frequently prevail in the atmosphere, and as a consequence the presence or absence of cloud formations is essentially due to favorable or unfavorable conditions of vertical velocity beneath the condensation level or the transition level (points A and B, respectively, in Figure 3.6).

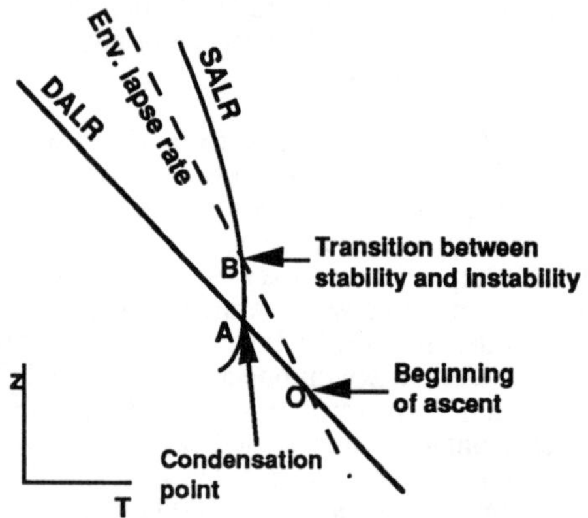

Figure 3.6: Ascent profiles in the presence of condensation

If, by analogy with the virtual temperature, a thermodynamic variable termed *virtual potential temperature* is defined as follows:

$$\theta_v = \theta (1 + 0.61 q) \tag{3.41}$$

then any stability states can be defined for all conditions of atmospheric moisture. If the atmosphere is dry, then $\theta_v = \theta$. If subscripts d refer to a dry atmosphere, v to a moist atmosphere, and w to saturated conditions, then the following stability conditions apply when considering the form of the vertical virtual potential temperatureprofile θ_v.

$$\frac{\partial\theta_w}{\partial z} = \frac{\partial\theta_d}{\partial z} \qquad \text{Absolute instability}$$

$$\frac{\partial\theta_w}{\partial z} < \frac{\partial\theta_v}{\partial z} < \frac{\partial\theta_d}{\partial z} \qquad \text{Conditional instability}$$

$$\frac{\partial\theta_v}{\partial z} = \frac{\partial\theta_d}{\partial z} \qquad \text{Neutral (dry atmosphere)}$$

$$\frac{\partial\theta_v}{\partial z} = \frac{\partial\theta_w}{\partial z} \qquad \text{Neutral (saturated atmosphere)}$$

$$\frac{\partial\theta_v}{\partial z} < \frac{\partial\theta_w}{\partial z} \qquad \text{Absolutely stable} \tag{3.42}$$

The term "absolute" indicates that the stability criteria are satisfied regardless of the prevailing humidity conditions in the atmosphere. These criteria are graphically summarized in Figure 3.7.

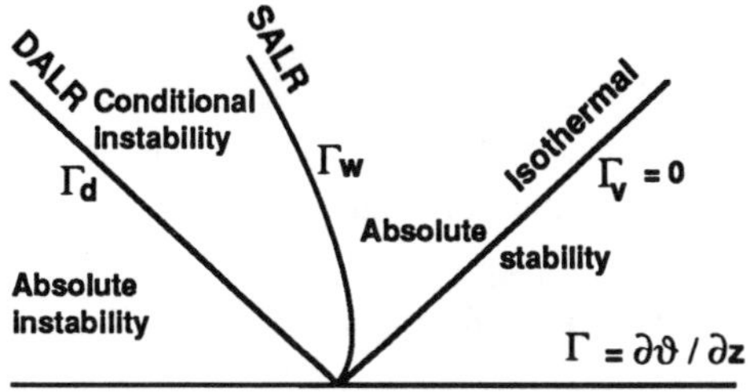

Figure 3.7: Stability criteria in a moist atmosphere

An example of particular interest in mountainous regions is given by the thermodynamic changes undergone by air masses during episodes of Föhn. Figure 3.8 illustrates the cycle as air rises by forced motion across a mountain barrier, condenses at a certain level and ascends further in the form of cloud. In the process of crossing the mountains, the clouds precipitate as rain (or snow) and therefore the air loses its moisture. On descending from higher levels on the leeward side of the mountains, the air will warm at the dry adiabatic rate since it no longer has any moisture and cannot descent at the saturated lapse rate. On arriving at low levels, the temperature elevation compared to that at the same altitude before ascent on the upwind side of the mountain barrier is considerable. Such extreme Föhn conditions, leading to compressional warming on the leeward side of the mountains, are a common feature of important mountain chains such as the Alps or the Rocky Mountains.

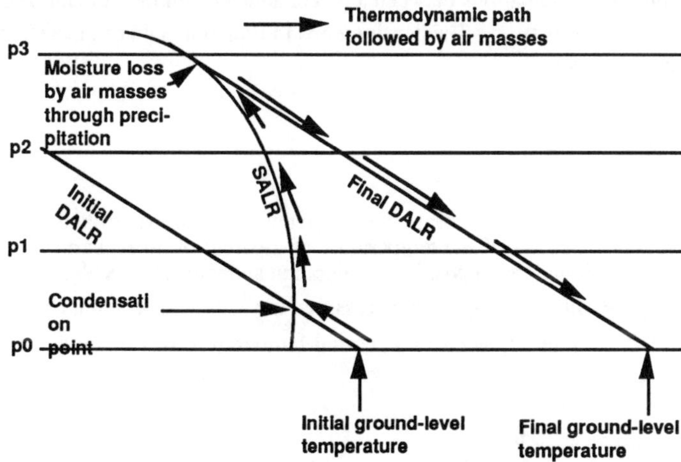

Figure 3.8. Thermodynamic description of the Föhn effect

3.2.5.2 The nature of vertical accelerations. Thermal forces which induce vertical motion in the atmosphere are known as *buoyancy* forces. Buoyancy in the atmosphere follows the well known Archimedes Principle where the buoyancy of a body is equal to the weight of fluid displaced by that body. This principle can be applied to a parcel of air with a weight given by:

$$W = \rho V g \tag{3.43}$$

where ρ is the density of the air parcel, V its volume, and g the gravitational acceleration. The weight of air displaced by the parcel is:

$$W' = \overline{\rho} \, V g \tag{3.44}$$

where $\overline{\rho}$ is the density of the air surrounding the parcel. Defining M as the parcel's mass and a its vertical acceleration leads to the resultant buoyancy force:

$$F = M a = \overline{\rho} \, V g - \rho V g = (\overline{\rho} - \rho) V g \tag{3.45}$$

However, since mass is density times volume, Equation 3.45 reduces to:

$$a = g \left(\frac{\overline{\rho} - \rho}{\rho} \right) \tag{3.46}$$

It is usually more convenient to express Equation 3.46 in terms of temperature, as this is more convenient to handle or to measure in the atmosphere than density. The equation of state yields:

$$\overline{\rho} = \frac{\overline{p}}{R \overline{T}} ; \qquad \rho = \frac{p}{R T}$$

However, pressure fluctuations between a parcel and surrounding air can be considered to be negligible, so that $p \cong \overline{p}$. The density terms in Equation 3.46 can therefore be replaced by temperature to give:

$$a = g \left(\frac{T - \overline{T}}{\overline{T}} \right) = g \left(\frac{T'}{\overline{T}} \right) \tag{3.47}$$

where the prime quantity represents the difference between parcel and air temperatures. Equation (3.47) states that if the temperature of a parcel of air is warmer than that of its environment, it will undergo a positive acceleration and rise. The acceleration will continue until $T' \leq 0$. Because parcel and air pressures are identical for equal pressure ambiant temperature, T may be replaced by potential temperature θ. Equation 3.47 can then be written as:

$$a = \frac{dw}{dt} = \frac{d^2z}{dt^2} = g \left(\frac{\theta - \overline{\theta}}{\overline{\theta}} \right) \tag{3.48}$$

If a parcel is initially at a temperature θ_0 and then is dispaced by a small increment δz, the environmental temperature may be written as:

$$\overline{\theta} \, (\delta z) = \theta_0 + \frac{d\theta}{dz} \delta z \tag{3.49}$$

If the temperature change is adiabatic, this implies that the potential temperature of the air parcel is conserved (i.e., $\theta \, (\delta z) = \theta_0$), so that the parcel/air temperature difference is:

$$\theta \, (\delta z) - \overline{\theta} \, (\delta z) = \frac{d\theta}{dz} \delta z \tag{3.50}$$

Using now Equation 3.48, the acceleration is given by:

$$\frac{d^2(\delta z)}{dt^2} = g\,(\frac{\theta - \bar{\theta}}{\theta}) = \frac{g}{\theta}\frac{d\theta}{dz}\,\delta z = -\,N^2\,\delta z \tag{3.51}$$

The parameter N $(= \sqrt{(\frac{g}{\theta}\frac{d\theta}{dz})})$ is a measure of the stability of the environment and is known as the *Brunt-Vaisala frequency*. Equation 3.51 has a general solution of the form:

$$\delta z = A\ \exp(i\,N\,t) \tag{3.52}$$

where A is the amplitude of the oscillation. If $N > 0$, the parcel will oscillate about its level of origin with a period given by:

$$\tau = \frac{2\pi}{N} \tag{3.53}$$

If $N^2 = 0$, this implies that $\frac{d\theta}{dz}$ is also zero, i.e., that the atmosphere is neutral, while if $N^2 < 0$ the displacement of the parcel will increase exponentially with time, characteristic of unstable conditions.

3.2.6 Stability considerations in the Atmospheric Boundary Layer (ABL)

In the absence of local perturbations such as those associated with cloud activity, the troposphere is generally characterized by absolute stability conditions above 1 or 2 km height; i.e., above the ABL. Within the lower atmospheric boundary layer, it was shown that surface-induced turbulence (cf. section 2.6) could modify dynamic quantities. It will be shown here that mechanical turbulence can also influence thermodynamic quantities, and that atmospheric stability can influence turbulence, thereby leading to a complex set of non-linear interactions.

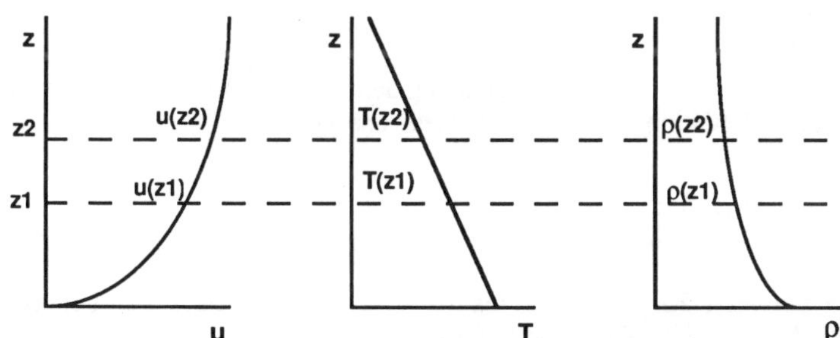

Figure 3.9: Typical profiles of u, T, and ρ in the Atmospheric Boundary Layer

Figure 3.9 illustrates typical profiles of wind, temperature, and density in the ABL. Since the length-scale l is defined as:

$$l = \delta z = z_2 - z_1 \tag{3.54}$$

the values of u, T and ρ at $z=z_2$ may be written as:

$$u(z_2) = u(z_1) + l \frac{\partial u}{\partial z}$$

$$T(z_2) = T(z_1) + l \frac{\partial T}{\partial z}$$

$$\rho(z_2) = \rho(z_1) + l \frac{\partial \rho}{\partial z} \tag{3.55}$$

Consider now a fluid element with volume V. Its total kinetic energy is composed of two independent factors:

- the horizontal displacement by the mean flow
- the turbulent motion due to dynamic and thermal effects.

The turbulent kinetic energy associated with vertical motions is given by:

$$KE_T = 1/2 \, \rho \, V \, w' \tag{3.56}$$

where w' represents the turbulent fluctuations of w. It may be recalled from mixing-length theory that for isotropic turbulence (u'=w') the fluctuating component of u may be defined as a function of $\frac{\partial u}{\partial z}$. According to simplified turbulence parameterization schemes, which will not be discussed here, the above equation may be re-written as:

$$KE_T = 1/2 \, \rho \, V \, l^2 \left(\frac{\partial u}{\partial z}\right)^2 \tag{3.57}$$

The work done by the parcel of air during its rising motion (work=force x distance) is given by:

$$W_p = 1/2 \, g \, (\rho(z_2) - \rho(z_1)) \, (z_2 - z_1) \, V \tag{3.58}$$

Substituting for $\rho(z_2) - \rho(z_1)$ and replacing density by temperature through the equation of state leads to definition of the vertical kinetic energy:

$$KE_v = -1/2 \, \rho \, V \, l^2 \, \frac{g}{T} \frac{\partial T}{\partial z} \tag{3.59}$$

The ratio between KE_v and KE_T results in a dimensionless number known as the *Richardson Number* Ri: . This is written as:

$$Ri = \frac{g}{T} \frac{\frac{\partial T}{\partial z}}{\left(\frac{\partial u}{\partial z}\right)^2} \tag{3.60}$$

or, in terms of potential temperature θ, as:

$$Ri = \frac{g}{\theta} \frac{\frac{\partial \theta}{\partial z}}{(\frac{\partial u}{\partial z})^2} \tag{3.61}$$

The Richardson Number represents the ratio between the dissipation or production of turbulent energy due to buoyancy effects and the mechanical turbulent kinetic energy. It states that if atmospheric conditions are stable, mechanical turbulence will be inhibited or suppressed by thermal stratification, while for unstable conditions ($Ri < 0$), thermal effects will combine with friction effects in order to enhance atmospheric turbulence.

In general, atmospheric turbulence is reduced or suppressed for $Ri \geq 0.25$. Beneath this critical value, turbulence will become mor important; during warm summer days, with considerable warming of the ground surface and a strong wind shear, Ri reach values lower than -1, leading to strong mechanical and thermal turbulence

3.2.7 Conservations equations for heat and moisture

It was seen earlier in this section that a change in potential temperature is equivalent to a change in entropy of the gas. The change in time of potential temperature of a parcel of air can be defined by taking the time derivative of Equation 3.15, i.e.:

$$\frac{c_p}{\theta}\frac{d\theta}{dt} = \frac{ds}{dt} = S_\theta \frac{c_p}{\theta} \tag{3.62}$$

where S_θ represents the sources and sinks of heat. From the chain rule of calculus, the Lagrangian or total derivative may be described as a function of its Eulerian derivative (fixed point in space). Equation (3.62) therefore reads:

$$\frac{d\theta}{dt} = \frac{\partial \theta}{\partial t} + u\frac{\partial \theta}{\partial x} + v\frac{\partial \theta}{\partial y} + w\frac{\partial \theta}{\partial z} = S_\theta \tag{3.63}$$

The major sources and sinks of temperature in the atmosphere are those related to phase changes of water, as seen in this section, to radiative flux exchanges (incoming solar radiation and emission and absorption of infrared radiation) which will be treated in the forthcoming section, and to dissipation of kinetic energy by mechanical and thermal turbulence. If these three processes are denoted respectively by C_θ, R_θ, and F_θ, then :

$$\theta = C_\theta + R_\theta + F_\theta \tag{3.64}$$

so that the conservation equation for heat, described by potential temperature, reads:

$$\frac{\partial \theta}{\partial t} = - u\frac{\partial \theta}{\partial x} - v\frac{\partial \theta}{\partial y} - w\frac{\partial \theta}{\partial z} + C_\theta + R_\theta + F_\theta$$

$$= - V . \nabla\theta + C_\theta + R_\theta + F_\theta \tag{3.65}$$

The conservation equation for moisture relates the total derivative of specific humidity q to the sources or sinks of humidity in the atmosphere, i.e.:

$$f(dq,dt) = S_q \tag{3.66}$$

The term S_q can be very complex, as it generally includes a combination of condensation, evaporation, crystalization, sublimation, and precipitation terms, in addition to turbulent transports of humidity. If these thermodynamic and turbulent processes are given by C_q and F_q, respectively, and the chain rule of calculus is applied to Equation 3.66, then the conservation equation for moisture is written as:

$$\frac{\partial q}{\partial t} = - u\frac{\partial q}{\partial x} - v\frac{\partial q}{\partial y} - w\frac{\partial q}{\partial z} + Cq + Fq$$

$$= - V \cdot \nabla q + Cq + Fq \qquad (3.67)$$

These are the equations which generally enter into a numerical model in one form or another, in order to be able to determine the time rate of change of temperature or moisture at every grid-point of a model. These equations then feed back into the dynamics, through pressure fluctuations induced by heat changes, themselves perturbed by radiative or convective processes. Any atmospheric model which attempts to couple both dynamics and thermodynamics becomes an exceedingly complex system to simulate and to interpret.

3.3 Atmospheric radiative transfer

It is the purpose of this section to provide a brief introduction to radiative transfer in the earth's atmosphere. The section will cover two aspects: First it introduces some basics of radiative transfer and develops the theory of radiative transfer to the extent that the computation of radiative heating rates (the term R_θ in Equations 3.64 and 3.65 is illustrated. Second, it gives a general account of the global radiation budget touching upon climate relevant issues like the greenhouse effect.

3.3.1 Some basics

Radiative transfer is a process whereby energy is transferred from one place to another without the necessity of a material medium. The nature of electromagnetic radiation can be explained by two models, the quantum theory and the wave theory. Quantum theory forms the basis for explaining how radiation is produced (emission and absorption) and also describes the interaction of radiation with materia, such as absorbing gases. The wave theory explains how electromagnetic radiation travels through media and how it is scattered at particles.

An isolated atmospheric molecule can only absorb and emit radiative energy in discrete amounts which correspond to three major types of molecular excitation: i) change of orbital states of the electrons ii) vibrational excitation due to a vibration of atoms relative to the center of mass of the molecule iii) rotational excitation due to molecule rotation about the center of mass

As a consequence of the discrete energy levels a spectral measurement of radiation emerging from matter typically exhibits a structure of absorption lines which in turn from ensembles of absorption bands. The high spectral variability of absorption and emission requires that realistic radiative transfer modelling needs a careful treatment of the integration over wavelength.

The wavelength of radiation is usually denoted by λ and measured in units of micrometers ($1\,\mu m = 10^{-6}$ m) or nanometer (1 nm $= 10^{-9}$m). Another measure frequently used is the number of waves per one centimeter which is called wavenumber (usually designated ν in cm^{-1} '). That is wavenumber is the inverse of wavelength.

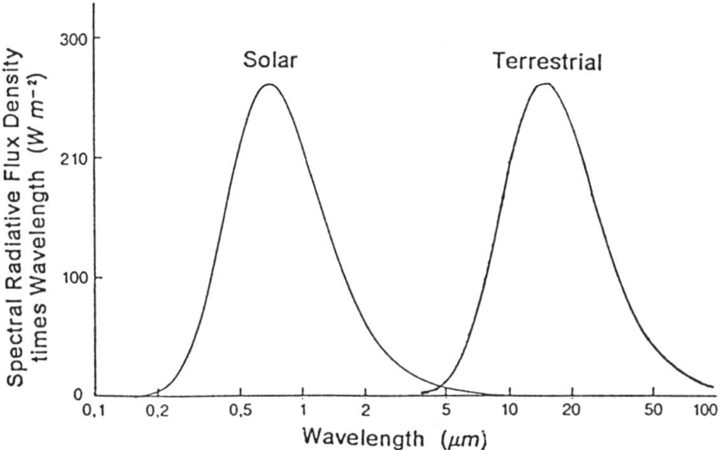

Figure (3.10): For the earth-atmosphere system a clear separation between terrestrial and soalr radition is possible for radiative energy studies, since both areas have little overlap. The axes are chosen such that areas are proportional to radiative energy. The solar radiation, equivalent to a black-body radiation for 5800 K, has been weighted with the squared ratio of the sun's radius over the mean distance between sun and earth. Another factor 0.25 accounts for the fact that the earth's cross section receives solar radiation while the whole sphere emits terrestrial radiation.

The wavelengths intervals of interest to radiative transfer in the earth's atmosphere comprise the longwave (4 -100 μm) spectrum, which covers radiation emitted within the earth's atmosphere, and the solar spectrum (0.2 - 4 μm) which covers the radiation received from the sun. The boundaries may look arbitrary but they are chosen such that the intervals encompass most of the total radiative energy which is of interest in energy budget studies (see Figure 3.10).

For most problems of atmospheric radiative transfer one is interested in time and space scales that are much larger than that corresponding to a photon or wavetrain. Therefore the radiation field is described by appropriate average quantities like:

- RADIANCE (denoted with the letter L; units $Wm^{-2}sr^{-1}$): radiative energy per time through an area into a solid angle $d\Omega$ (see Figure 3.11). This is the quantity a narrow-beam radiometer, for instance on a satellite, would observe.
- RADIATIVE FLUX DENSITY (denoted with the letter *F;* units Wm^{-2}) the radiative energy per time and area. Often the radiative flux density is just called radiative flux. This is the quantity which is of interest in energy budget studies.

Note that radiances and fluxes are sometimes spectrally defined (i.e. per wavelength). Then the energy related values need to be obtained by spectral integration.

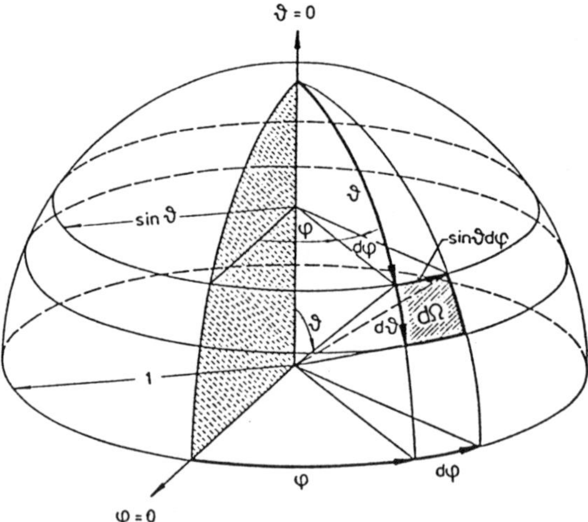

Figure (3.11) Schematic illustrating the definition of the solid angle dQ, the radiance and the radiative flux density (see Equation 3.68).

The radiative flux density F is computed from a radiance field L by integration over a hemisphere:

$$F = \int_0^{2\pi} \int_0^{\pi/2} L(\theta,\phi)\cos\theta\, d\Omega = \int_0^{2\pi} \int_0^{\pi/2} L(\theta,\phi)\cos\theta \sin\theta\, d\theta\, d\phi \tag{3.68}$$

where θ is the zenith angle and ϕ denotes the azimuth angle. Using the average quantities, like radiance and radiative flux density, implies that the elementary photon processes are treated in a parameterized from via absorption coefficients, scattering functions and the solar intensity or the Planck function as source functions.

3.3.2 Radiative transfer equation

When an incident spectral radiance L_λ passes through an absorbing or scattering medium over a path *ds* the incident radiance will be lessened. The amount of radiation dL_λ removed is proportional to the radiance Ll_λ and the length of the path *ds:*

$$dL_\lambda = -\sigma_{ex_1\lambda} L_\lambda ds \tag{3.69}$$

where $\sigma_{ex,\lambda} = \sigma_{ab,\lambda} + \sigma_{sc,\lambda}$ is the volume extinction coefficient (unit: m^{-1}) which is the sum of the volume absorption and volume scattering coefficients. Equation (3.69) is known as Beer's law which is valid for monochromatic radiation. In the following we will omit the subscript λ, which denotes the monochromatic radiation, for the sake of the simplicity.

In a horizontally homogeneous atmosphere it is useful to refer to the vertical coordinate z (altitude) and define the vertical optical depth δ as:

$$d\delta = \sigma_{ex} dz = \sigma_{ex} ds \mu \tag{3.70}$$

where $\mu = \cos\theta$, with θ being the angle between the path s and the vertical coordinate z.

While the radiation travels the distance $ds = dz / \mu$ an amount dJ is added to the beam due to emission and/or scattering. Adding this term to Equation (3.69) and taking the optical depth δ as variable describing the path one obtains:

$$\mu \frac{dL(\delta)}{d\delta} = -L(\delta) + J(\delta) \tag{3.71}$$

The above equation can be directly integrated after multiplication with an integrating factor of exp(δ / μ):

$$\mu \frac{dL(\delta)}{d\delta} \exp(\delta / \mu) + L(\delta)\exp(\delta / \mu) = J(\delta)\exp(\delta / \mu) \tag{3.72}$$

Integration over the path from O to δ' gives:

$$[L(\delta)\exp(\delta / \mu)]_0^{\delta'} = \int_0^{\delta'} J(\delta)\exp(\delta / \mu) d\delta \tag{3.73}$$

$$L(\delta')\exp(\delta' / \mu) - L(0) = \int_0^{\delta'} J(\delta)\exp(\delta / \mu) d\delta \tag{3.74}$$

$$L(\delta') = L(0)\exp(-\delta' / \mu) + \int_0^{\delta'} J(\delta)\exp(-(\delta' - \delta) / \mu) d\delta \tag{3.75}$$

The terms in Equation (3.75) have a simple physical meaning. The term on the left side is the radiance at the height level corresponding to the optical depth δ'. The first term on the right side is the radiance that emerges at δ = O and is transmitted over the path δ / μ. The part $J(\delta)d\delta$ of the integral is the emission over the optical path $d\delta$, that is multiplied with the transmission over the path from δ to δ'. Since all parts of the total path can contribute to the radiance $L(\delta')$ one has to integrate over the total path from O to δ'.

The term $J(\delta)$ is called the source function. The name is already self-suggestive. Generally it consists of two parts, namely thermal emission and scattering. Assuming local thermodynamic equilibrium (LTE) the emission part of the source function is described by the Planck function B which may vary over the optical depth δ:

$$J(\delta)d\delta = B(\delta)d\delta \tag{3.76}$$

The LTE assumption implies that the excited states of the absorbing gas are chiefly determined by collisional rather than radiative processes. That is, the time of excitation is much longer than the average time between collisions. LTE is fulfilled throughout the troposphere and lower startosphere.

The scattering component of the source function is due to the scattering of radiances $L(\theta';\phi')$ from all directions of local zenith angles θ' and azimuth angles ϕ' into the direction $(\theta;\phi)$ of the radiance under consideration. For solar radiation the source function is separated into the contribution from the direct solar beam and the contribution from the diffuse radiation field.

3.3.3 Heating rates

For the atmospheric heating rate calculations the required quantity are the radiative flux densities into the upper (F^+) and lower hemispheres (F^-), respectively. The integration over the hemispheres is conducted according to Equation 3.68. Since all wavelengths contribute to the fluxes an integration of the monochromatic fluxes with respect to wavelength must be performed. The broadband fluxes F^+ and F^- as functions of the vertical coordinate z are calculated from the spectral fluxes by:

$$F^{+-}(z) = \int_0^\infty F_\lambda^{+-}(z)\,d\lambda \tag{3.77}$$

The net flux at an altitude z is the diflerence between the upward and downward fluxes:

$$F_{net}(z) = F^+(z) - F^-(z) \tag{3.78}$$

The divergence or convergence of the net fluxes determines the local cooling or heating rates:

$$R_\theta = \frac{\partial T}{\partial t} = -\frac{1}{\rho c_p}\frac{dF_{net}(z)}{dz} \tag{3.79}$$

which is the diabatic heating term in Equation 3.65 due to the radiative flux divergence. As already discussed above it is useful to distinguish between infrared thermal radiation and solar radiative transfer. Thus in practise the two components are treated by different parametrizations. Example results for the heating rates due to infrared and solar radiation in a cloudy atmosphere are presented in Figure 3.12.

3.4 Clouds and the Earth's radiation budget

3.4.1 The Earth's radiation budget

The geographical distribution of the regional radiation budget of the earthatmosphere system is essential for maintaining the atmospheric and oceanic general circulation, since it constitutes the sources and sinks of energy for the Earthatmosphere system. The radiation balance is, therefore, a key quantity in climate research. Averaged over the year and the globe the absorbed solar radiative flux is:

$$\bar{S} = \pi r_e^2\, S_0 (1 - \alpha_{toa}) \tag{3.80}$$

where re is the Earth's radius, S_0 the mean solar constant $(\approx 1372 Wm^{-2})$, that is the radiative flux density for the mean distance between sun and Earth, and α_{toa} is the planetary albedo (about 0.3).

The solar energy input is counterbalanced by the longwave radiative loss of the Earth:

$$\bar{E} = 4\pi r_e^2\, \sigma T_e^4 \tag{3.81}$$

where σ is the Stephan-Boltzmann constant and T_e the mean radiative temperature of the Earth.

Results for the global earth radiation budget have been acqired over the last 25 years from satellite with increasing accuracy and success. Reviews are provided by Stephens et al. (1981) and Kandel (1990). The main points concerning the earth radiation budget at the top of the atmsophere are:

- the global annual mean albedo of the earth-atmosphere system is close to 0.3. The range of the annual variation is between 0.27 and 0.31
- the corresponding longwave radiative flux to space is close to the mean solar gain of 240 Wm^{-2} Its annual cycle is dominated by the larger range of the amplitude in the Northern hemisphere due to the continental distribution
- the global mean radiation balance (net radiation) goes through an annual cycle with an amplitude of $+/-11 \quad Wm^{-2}$ about the annual mean which not distinguishable from zero. This cycle is due to the storage of solar energy absorbed by the Southern oceans during austral summer when the Earth is closest to the sun

Equating 3.80 and 3.81 and putting in the value 0.3 for the planetary albedo yields the mean earth radiative temperature T_e:

$$T_e = \left[\frac{S_0(1-\alpha_{toa})}{4\sigma} \right]^{\frac{1}{4}} \approx 255K \tag{3.82}$$

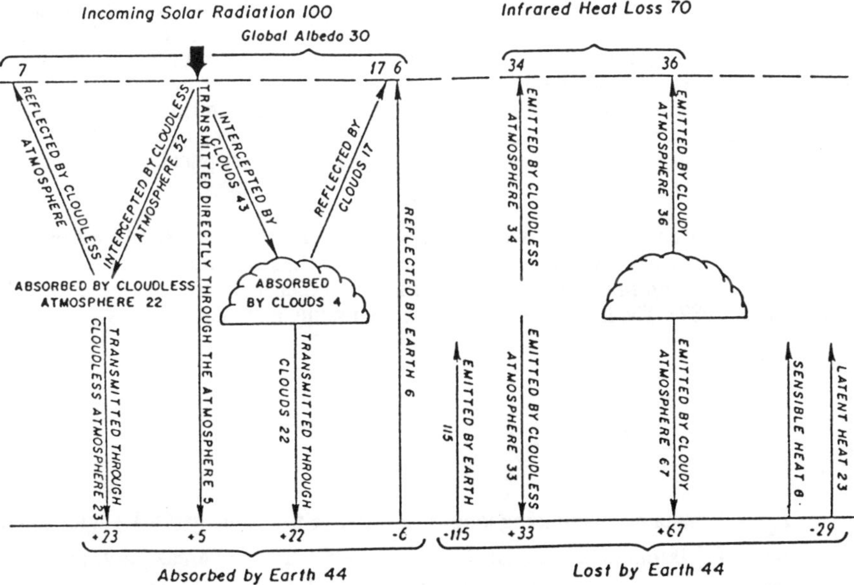

Figure (3.12): The energy balance of the earth-atmosphere system. The solar input is set to 100 units (from Liou, 1992).

A temperature of 255 K is representative for the atmosphere at an altitude of about 5 km. However the mean surface temperature is with about 288 K considerably higher. This is explained by the greenhouse trapping or greenhouse effect of the earth's atmosphere, where absorbing gases, clouds and aerosol absorb the upwelling longwave radiation from lower levels and emit at a lower temperature.

A detailed picture of the global annual radiation and heat budget of the earthatmosphere system is shown in Figure 3.12 (from Liou, 1992). The incoming solar radiation is set to 100 units. 44 units of the solar input are absorbed at the earth's surface while 26 units are absorbed by the atmosphere; 4 by clouds and 22 by the clear atmosphere. The net longwave flux at the surface is -15 units which does not fully balance the solar absorption. The balance is maintained by the surface losses due to latent and sensible heat. Since the annual mean of the ratio of sensible to latent heat loss at the surface (the Bowen ratio) is about 0.27, the partitioning between sensible and latent heat is 6 units and 23 units, respectively. As an important point one can state that the solar heating is mainly takes place at the surface while the longwave cooling mainly occurs in the atmosphere.

3.4.2 Radiation budget and the role of clouds

The role of clouds is one of the major uncertainties in the evaluation of the sensitvity of the earth's climate to perturbations. The effect of clouds on the outgoing longwave radiation field and the absorbed solar radiation are both potentially large and have opposing effects on the net radiation (Schneider, 1972; Raschke, 1993; Rossow, 1993). In the solar shortwave spectrum clouds generally increase the planetary albedo and reduce the solar irradiance at the surface. In this respect clouds have a cooling effect which is primarily felt at the surface. The influence on the atmosphere depends on cloud height, that is low cloud increase the absorption in the atmosphere in comparison to clear sky while high clouds reduce the solar heating of the atmosperic column. In the thermal longwave spectrum clouds tend to increase the atmospheric greenhouse effect, which is most pronounced for high clouds.

An important issue in the current debate on climate change is the question whether a change in cloud associated with a change in climate amplifies (positive feedback) or reduces (negative feedback) the initial forcing of a temperature change. As it turns out clouds could do both. For instance, if a temperature increase due to an external forcing is associated with an increase in low level cloud which are bright and warm, then this would enhance the planetary albedo more than it would affect the greenhouse trapping of the atmosphere; thus the feedback would be negative. The opposite is conceivable for optically thin and cold high level cloud (cirrus).

Obviously a feedback as such is not an observable quantity. A useful and observable quantity that describes the direct eflect of clouds on net the radiation at the top of the atmosphere is commonly termed 'cloud forcing' (Charlock and Ramanathan, 1985; Ramanathan, 1987). The longwave cloud radiative forcing may be written as:

$$CF_{lw} = F_{cl} - F_a \tag{3.83}$$

or:

$$CF_{lw} = C(F_{cl} - F_{ov}) \tag{3.84}$$

where F_a is the area mean outgoing longwave radiative flux density (OLR) at the top of the atmosphere, F_{cl} is the clear sky OLR, F_{ov} is the OLR for overcast sky, and C is the fractional cloud cover. Equations similar to 3.83 and 3.84 can be written for the solar shortwave cloud forcing.

The concept of cloud-radiative forcing is appealing for its apparent simplicity as it does not require the explicit estimation of cloud parameters from satellite data when the total forcing is considered. By estimating the total forcing CF_{lw} the problem reduces mainly to the retrieval of the clear sky fluxes F_{cl} and the inference of an area mean flux including clouds.

Observing the global distribution of the cloud-radiative forcing and its seasonal change is a first step toward understanding the cloud-radiative feedback to climate perturbations. Overviews of the topic are given by Ramanathan (1987) and Ramanathan et al. (1989); the latter paper also presents first results of the global shortwave and longwave components of the cloud forcing from ERBE (Earth Radiation Budget Experiment; Barkstrom and Smith, 1986).

Concerning the present knowledge on the role of clouds it is worthwhile to mention the outstanding result of the Earth Radiation Budget Experiment (ERBE): The analysis by Ramanathan et al. (1989) shows: first, the global mean cloud forcing components in the longwave and shortwave are nearly an order of magnitude larger than the radiative forcing from an instantaneous CO_2 doubling (about 4 Wm^{-2}) and second the net effect of the present cloud distribution appears to be a cooling of the earth, that is the albedo eflect of clouds dominates over their greenhouse effect.

3.4.3 Surface radiation budget

The shortwave radiation budget at the surface is determined by the downward shortwave flux $S\downarrow$ and the surface albedo α_{sfc}; the longwave budget requires knowledge of the downwelling longwave flux $L\downarrow$, the surface temperature T_{sfc} and the surface emissivity ε_{sfc}. The total surface radiation budget can be written as:

$$R_{net} = S\downarrow(1-\alpha_{sfc}) + L\downarrow - \varepsilon_{sfc}\sigma T_{sfc}{}^4 \tag{3.85}$$

The shortwave balance is always positive, while in the longwave budget the surface exitance generally exceeds $L\downarrow$. The shortwave surface flux is closely coupled to the flux observed at the top of the atmosphere which explains why is possible to directly estimate the components of the surface solar radiation field from satellite (Chou, 1991; Li et al., 1993). Figure 3.12 reveals the reason for this, namely that the atmosphere is fairly transparent to solar radiation.

Solar absorption in the atmosphere varies with wavelength and with solar zenith angle. Ozone and molecular oxygen nearly completely absorb the radiation below 0.3 ,μm. At wavelengths larger than about 0.7 μm water vapour absorption becomes important. Rayleigh scattering by molecules dominates the extinction in the visible part of the solar spectrum; it rapidly decreases with wavelength since it is proportional to λ^{-4}. Scattering by particles, called Mie-scattering, has a much weaker dependence on wavelength ($\approx \lambda^{-1.3}$). Cloud absorption depends on the cloud liquid or ice water content and on the droplet or particle size distribution. Cloud altitude is also important; generally one can say that for a given atmosphere the solar absorption by the atmosphere increases as the cloud altitude is lowered (e.g. Schmetz, 1993).

The surface albedo is defined as the ratio of the broadband reflected $(S\uparrow)$ over the incident solar flux density $(S\downarrow)$ at the surface:

$$\alpha_{sfc} = \frac{S\uparrow}{S\downarrow} \tag{3.86}$$

It is important to note that the surface albedo is not an inherent property of the surface. For natural surfaces α_{sfc} generally depends on the angular and spectral distribution of the irradiance $S\downarrow$. The truly inherent surface property is the spectral reflection function $\lambda_\lambda(\mu_i,\phi_i;\mu_r,\phi_r)$ which determines what fraction of the monochromatic radiance incident from the direction (μ_i,ϕ_i) is being reflected into the direction (μ_r,ϕ_r). For instance, Kriebel (1979) has shown that the albedo of vegetated surfaces may vary by more than 2% (absolute) with solar zenith angle. This occurs because the spectral distribution of the irradiance varies with zenith angle.

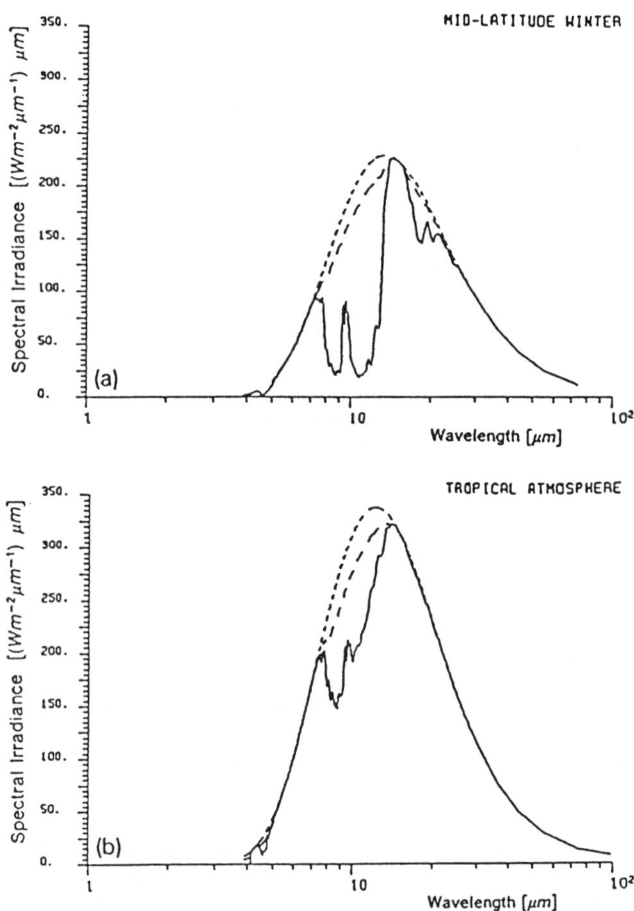

Figure (3.13): Spectral longwave downward longwave fluxes at the surface vs. wavelength for two different atmospheres characteristic of a mid-latitude winter and the tropics, respectively. Abscissa and ordinate are such that area corresponds to radiative energy. The solid line corresponds to clear sky and long dashes to a cloud base at 2 km altitude. The short dashes give the Planck function at surface temperature.

The downward longwave irradiance (4 -100 μm) at the surface is the result of atmospheric absorption, emission and scattering of the entire atmospheric column and it depends on the vertical profiles of temperature, gaseous absorbers and clouds. Effectively, however, $L\downarrow$ is determined by the radiation emitted in a shallow layer close to the surface (Schmetz et al.,

1986). Contributions from above 500 meters amount to only about 20% (midlatitude winter) and 16% (tropical), respectively. The lowest 10 meters of the atmosphere already emit 32% and 36% of the total surface fluxes of 222 Wm^{-2} and 407 Wm^{-2}.

Cloud contributions are mainly from the atmospheric window region (8 -13 μm) and the relevant parameters are cloud base height and temperature and cloud cover. The relative importance of cloud contributions decreases within moister climates since the transparency of the window decreases due to water vapour continuum absorption. Figure 3.13 shows the spectral downward longwave flux at the surface multiplied with the wavelength versus the logarithm of the wavelength for two standard atmospheres (Schmetz, 1989). The solid lines depict the clear sky spectral irradiance, short dashes denote the Planck function for surface temperature, and the long dashes show the irradiance from a sky with a black cloud at 2km altitude (cloud base). It is evident that the cloud contribution is confined mainly to the atmospheric window region between 8 and 13 μm. The cloud contribution in the midlatitude winter atmosphere is 74.5 Wm^{-2} or 33.6%, while in the tropical atmosphere the contribution is only 33.5 Wm^{-2} or 8.2%.

As concluding remark we can summarize that the net longwave radiation at the surface is determined by the surface temperature and emissivity and the near-surface temperature and humidity fields. Upper levels of the atmosphere contribute only through clouds in the atmospheric window region. This is why the longwave downward flux can be estimated so well from screen temperature (at 2m) via empirical formulae.

3.5 Effects of radiation on clouds

The feedbacks between radiation and cloud processes is a major theme in current climate research, in particular because the sign of the feedback of clouds on the climate system can be both negative or positive. In curent climate models, the grid structure does not resolve clouds and their parameterization is often inadequate, so that the net effect of clouds on climate is difficult to determine on the basis of such models. Mesoscale model studies can help alleviate this problem by providing a resolution which is sufficient for process interaction studies, thereby highlighting the mechanisms involved and providing a means to improve parameterizations in models with lower spatial resolution.

Beniston and Schmetz (1985) and Schmetz and Beniston (1986) used a three-dimensional mesoscale atmospheric model to investigate the influence of infrared and solar radiation schemes on model clouds; this section will summarize the results of those studies to illustrate the sensitivity of one feature of the mesoscale atmosphere to forcings from another component of the system.

The study made use of data from an amospheric experiment over the North Sea conducted in 1981 (the KonTur - Convection and Turbulence Experiment; see Hoeber, 1982) by several German and international groups. Based on the available data, profiles of wind, temperature and moisture typical of a cool maritime boundary layer were provided for the initialization of the mesoscale model. Conditions over the North Sea were favorable for the formation of shallow to medium-depth cumulus clouds, characteristic of Rayleigh-Bénard type convection where cool air overlies a warmer ocean surface (see Beniston, 1985, for example). The aim of the coupled cloud-radiation simulations was therefore to determine the extent to which the activity of the shallow cumulus clouds is sensitive to both infrared and infrared + solar radiative forcing. In order to achieve this aim, a control experiment was carried out with the mesoscale model to provide reference data for the case with only prescribed 'radiation cooling', i.e., a cooling of 1 K/day imposed at all grid-points. Using an infrared parameterization scheme, the simulations were repeated with all other initial and boundary conditions identical. The infrared scheme was calculated in two manners, i.e., either a computation of the contribution of infrared cooling directly at each grid point, or a smoothing of the influence of the radiative forcing over adjacent grid points at different levels; this latter simulation was carried out to remove possible spurious effects imposed by the relatively coarse vertical grid spacing which was used in the mesoscale model at that time.

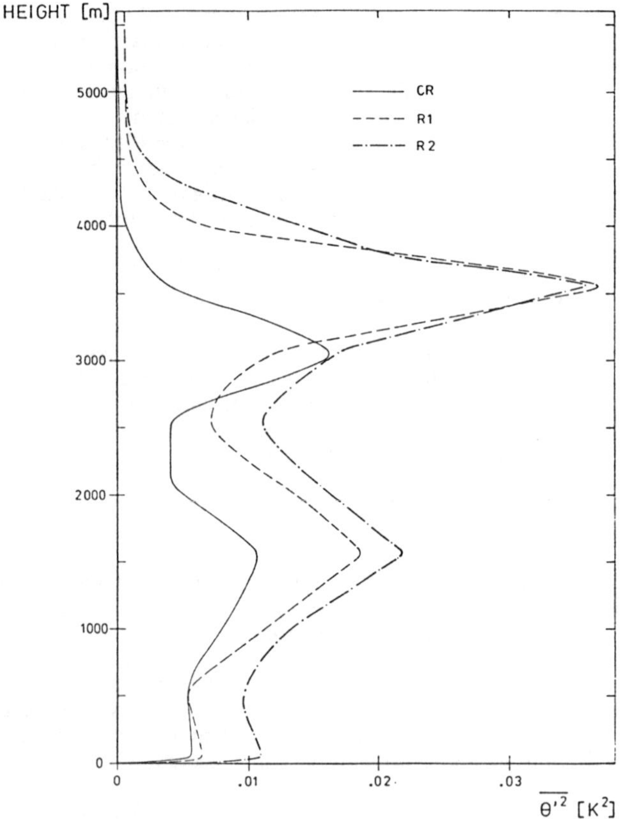

Figure 3.14: Variance of temperature in a cloudy air column for mesoscale simulations with and without infrared parameterization scheme (reproduced from Boundary-Layer Meteorology, 1985, courtesy of Kluwer Academic Publishers, The Netherlands)

Figure 3.14 illustrates the variance of temperature in the model after 12 hours of simulation, at which time an active field of cumulus clouds has developed to the mature phase. The signals of cloud-base condensation warming and strong cloud-top evaporational cooling are quite clearly identified in this figure; CR denotes the control experiment, and R1 and R2 the simulations with radiative forcing (applied at each grid point, and smoothed over vertical levels, respectively). It is observed that the warming and cooling features in the cloud layer are enhanced by the infrared forcing; it is also observed that the region of maximum cloud-top cooling is shifted to higher levels with respect to the control experiment CR by about 500 m. This is because the cloud top cooling enhanced by infrared forcing tends to destabilize to a greater degree the atmosphere at these levels, allowing additional cloud extent in the vertical to occur; as a result, the region in which cloud-top detrainment of liquid droplets and their evaporation into the cloud-free environment takes place at higher elevations. This enhanced destabilization of the atmosphere allows the clouds to reach their mature stage faster than in the case without infrared cooling (after 7 hours in R1 and R2 compared to 10 hours in CR). The mass fluxes are doubled in R1, R2 with respect to CR, because of the larger quantitites of air circulating within the thicker cloud layers resulting from the enhanced cloud activity. This implies that the compensating subsidence in the cloud-free atmosphere will also be stronger (in order to satisfy considerations related to mass continuity), so that the cloud-free air will tend to be drier and warmer than in the control experiment with no infrared scheme.

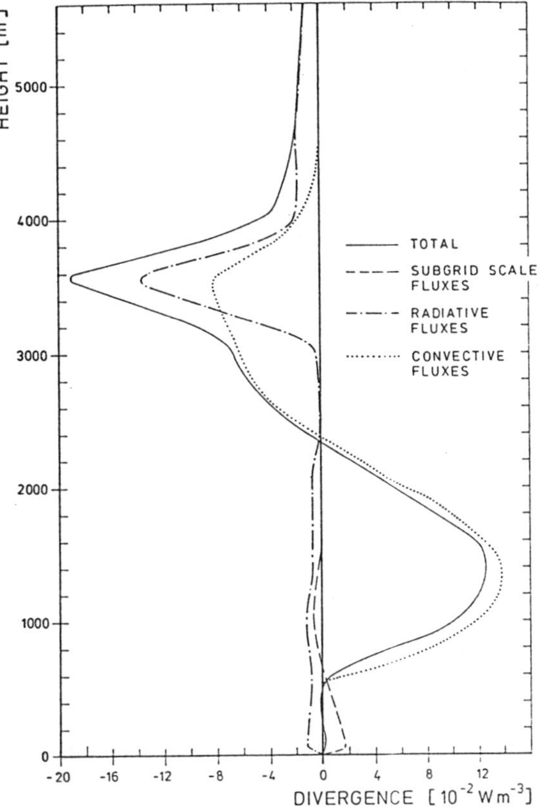

Figure 3.15: Contributions of thermodynamics, radiation, and turbulence to heat budgets in the mesoscale model in the presence of clouds (reproduced from Boundary-Layer Meteorology, 1985, courtesy of Kluwer Academic Publishers, The Netherlands)

Figure 3.15 illutrates the contributions of the different subgrid, thermodynamic, and radiative terms which make up the thermodynamic equation defined by Equation 3.65, i.e., of C_θ , R_θ , and F_θ. Radiative cooling is substantially stronger than evaporation cooling, but the two combine to produce an intense heat sink close to cloud tops. At cloud base, condensation warming is the dominant term and radiative cooling is minor; the result, as previously mentioned, is a locally unstable, very turbulent situation within the cloudy environment. The amount of rainwater generated by the clouds is 10 - 20% higher in the situation with the infrared scheme than in CR, which provides an additional confirmation of the enhanced activity generated by the radiative forcing.

When solar radiation is included as an additional component of the R_θ term in Equation 3.65, the resulting situation is somewhat different. Using the same procedures and conditions for the numerical simulations, the solar component of radiative forcing tends to reduce the inensity of infrared cooling. In the experiments reported by Beniston and Schmetz (1986), it was seen that for a chosen zenith angle of 60°, the magnitude of the solar radiation was 40% of that of infrared radiation, but of opposite sign. This is illustrated in Figure 3.16, where the solar, infrared, and net cooling curves are given for a column in which clouds are present. This has a particular consequence for the intensity of cloud-top cooling, which is reduced in the case of solar + infrared forcings as compared to the infrared cooling case alone. The results for cloud activity (cloud-top height, rainwater production, vertical mass flux, etc.) are therefore somewhat intermediate between the CR experiment and the R1, R2 numerical studies in which only infrared forcing is taken into account.

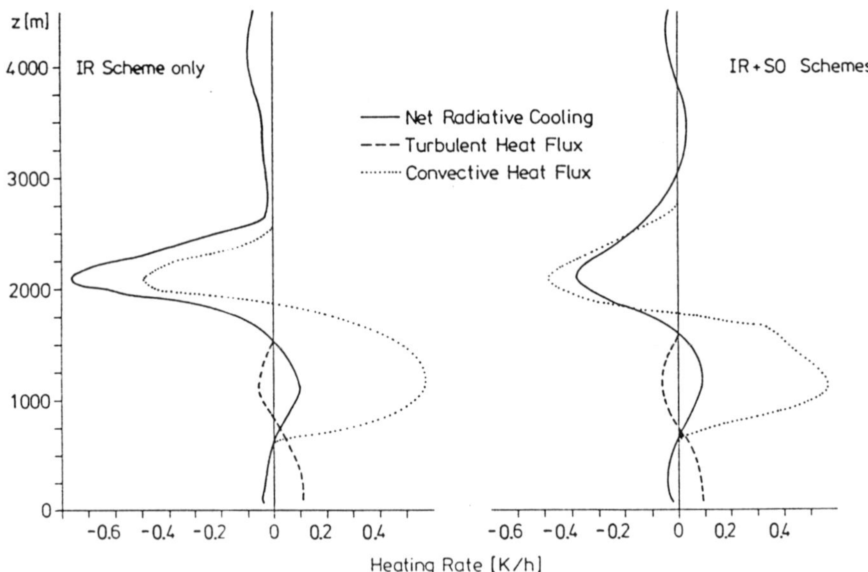

Figure 3.16: Contributions of solar and infrared terms to heat budget in the mesoscale model for a cloudy air column (reproduced from Boundary-Layer Meteorology, 1986, courtesy of Kluwer Academic Publishers, The Netherlands)

These studies have underlined the need to adequately parameterize radiation physics in an atmospheric model, both on the regional scale in order to determine the important underlying processes, and on larger (hemispheric or global) scales so that the thermodynamic and radiative signatures of clouds on the climate system are more coherently represented than in present climate models. The mesoscale simulations summarized here have shown that even if solar radiation modulates to some extent the magnitude of radiative cooling, the impact of the radiative scheme on cloud features is greater on cloud activity than when no scheme is incorporated into the model. Indeed, if clouds were to persist during nocturnal conditions, then the infrared cooling as in the R1, R2 experiments would represent the dominant thermal forcing term, resulting in an intensification of cumulus activity in the type of boundary-layer considered here.

3.6 Conclusions

This chapter has covered some of the fundamentals of atmospheric thermodynamics and radiation. The First and Second Laws of Thermodynamics have been introduced, and their applications to atmospheric processes demonstrated; in particular, the behavior of the atmosphere in the presence of phase chages of water makes use of the basic principles of thermodynamics to describe the complexity of condensation, evaporation, and convection. The chapter has also presented some of the basics of atmospheric radiative transfer and radiative heating rates in order to familiarize the reader with another aspect of atmospheric physics which is of fundamental importance to meteorology and climatology on all scales of the atmospheric spectrum; in particular, the ongoing debate on global atmospheric warming, through the increased concentrations of greenhouse gases of anthropogenic origin, focuses upon the radiative properties of these gases and their potential for warming the atmosphere.

In order to illustrate that atmospheric physics is not a set of physical mechanisms rigidly separated by discipline (i.e., dynamics, thermodynamics, radiation, etc.), the interaction between two major elements of the climate system, namely clouds and radiation, and its effect on the earth's radiation budget, has been described on a theoretical basis in Section 4. The role of clouds and their influence on radiation, and hence on atmospheric warming through negative or positive feedbacks (i.e., cooling through reflection of solar energy or warming through back-radiation of outgoing terrestrial radiation) have been covered.

Finally, as a specific example of these mechanisms, an overview has been given of some of the past work of the authors on the numerical simulation of cloud-radiation interactions, using a mesoscale atmospheric model which incorporates to a sufficient degree of detail the physics and dynamics of the system, thereby enabling meaningful results to be obtained.

3.7 References

Barkstrom, B.R. and G.L. Smith, 1986: The earth radiation budget experiment: Science and implementation. Rev. Geophys., 24, 379 - 390.

Beniston M., 1985 : Organization of convection in response to initial and lower boundary conditions in a mesoscale numerical model. Contr. Atm. Phys., 58, 31-52.

Beniston, M., and J. Schmetz, 1985: A three-dimensional study of mesoscale model response to radiative fircing. Boundary-Layer Meteorol., 31, 149 - 175

Cess, R.D. and Vulis, I.L., 1989. Inferring surface albedo solar absorption from broadband satellite measurements. J. Climate, 2: 974 - 985.

Charlock, T.P. and V. Ramanathan, 1985: The albedo field and cloud radiative forcing produced by a general circulation model with internally generated cloud optics, J. Atmos. Sci., 42, 1408-1429.

Chou, M.-D., 1991: The derivation of cloud parameters from satellite- measured radiances for use in surface radiation calculations. J. Atmos. Sci., 48, 1549 1 559.

Hoeber, H. (Ed.), 1982: KonTur: Convection and Turbulence Experiment. Hamburg Geophys. Monogr., 89 pp.

Holton, J. A., 1972: An Introduction to Dynamic Meteorology. Academic Press, New York, 319 pp.

Kandel, R.S. 1990. Satellite observation of the earth radiation budget and clouds. Space Science Reviews, 52, 1 - 32.

Kriebel, K.T., 1979. Albedo of vegetated surfaces: Its variability with differing irradiances. Remote Sens. Environm., 8: 283 - 290.

Li, Z., H.G. Leighton, K. Masuda and T. Takashima, 1993: Estimation of SW flux absorbed at the surface from TOA reflected flux. J. Climate, 6, 317 - 330.

Liou, K.N., 1992: Radiation and cloud processes in the atmosphere. Oxford Monographs on Geology and Geophysics No. 20, Oxford University Press, pp. 487.

Manabe, S. and R.T. Wetherald, 1967: Thermal equilibrium of the atmosphere with a given distribution of relative humidity. J. Atmos. Sci., 24, 241 - 259.

Pielke, R. A., 1984: Mesoscale Meteorological Modeling. Academic Press, New York, 612 pp.

Ramanathan, V., 1987: The role of earth radiation budget studies in climate and general circulation research. J. Geophys. Res., 92, D4, 4075 - 4095.

Ramanathan, V., R.D. Cess, E.F. Harrison, P. Minnis, B.R. Barkstrom, E. Ahmad, and D. Hartmann, 1989: Cloud-radiative forcing and climate: Results from the earth radiation budget experiment. Science, 243, 57 - 63.

Raschke, E., 1993: Radiation-cloud-climate interaction. NATO ASI Series, Vol. 1 5, Energy and Water Cycles in the Climate System, Eds. E. Raschke and D. Jacob, Springer-Verlag, p. 69 - 93.

Rossow, W.B., 1993: Satellite observations of radiation and clouds to diagnose energy exchanges in the climate: Part I + 11, NATO ASI Series, Vol. 15, Energy and Water Cycles in the Climate System, Eds. E. Raschke and D. Jacob, Springer-Verlag, p. 123-164.

Schmetz, J., and M. Beniston, 1986: Relative effects of solar and infrared radiative forcing in a mesoscale model. Boundary-Layer Meteorol., 34, 137 - 155
Schmetz, P., J. Schmetz and E. Raschke, 1986: Estimation of daytime downward longwave radiation at the surface from satellite and grid point data. Theor. Appl. Climatol., 37, 136-149.
Schmetz, J., 1989: Towards a surface radiation climatology: Retrieval of downward irradiances from satellites. Atmospheric Research, 23, 287 - 312.
Schmetz, J., 1993: Relationship between solar net radiation at the top of the atmosphere and at the surface. J. Atmos. Sci., 50, 1122 -1132.
Schneider, S.H., 1972: Cloudiness as a global feedback mechanism: effects on the radiation balance and surface temperature of variations in cloudiness. J. Atmos. Sci., 29, 1413-1422.
Stephens, G.L., G.G. Campbell and T.H. vonder Haar, 1981: Earth radiation budgets. J. Geophys. Res., 86, 9739 - 9760.

Authors's affiliations:

1: Department of Geography, ETH-Zürich, Switzerland
2: ESA - European Space Operations Centre, Darmstadt, Germany

IV Atmospheric Boundary-Layer Processes and Influence of Inhomogeneous Terrain

F.T.M. Nieuwstadt

4.1 Introduction

The boundary layer is defined as the lowest part of the atmosphere. In it meteorological variables, such as wind velocity, temperature and humidity, adjust from their values in the free-atmosphere to the boundary conditions at the earth's surface. These boundary conditions are the no-slip condition for the velocity and for heat and humidity imposed surface fluxes. The theory of the atmospheric boundary layer should provide us with the vertical profiles of these meteorological variables and their fluxes. It will be clear these profiles will depend strongly on the process that determines vertical transport in the boundary layer. This process is turbulence. Therefore, the study of the atmospheric boundary layer is almost synonym with a study of atmospheric turbulence.

In a description of atmospheric turbulence we can recognize various categories. These categories are classified according to the intensity and structure of turbulence. In particular, we will find that each category can be characterized by dimensionless parameters which are defined in terms of the production processes and also in terms of the length and time scale of turbulence. With these parameters we can then identify so-called scaling regimes of the atmospheric boundary layer. In these scaling regimes some physical process usually dominates. This allows us to introduce some simplifications or approximations in our description of the boundary layer. As a result we may distinguish so-called prototypes of the atmospheric boundary, which are representative for some meteorological condition. These prototypes, which we shall discuss in detail in the following sections, are: the surface layer, the neutral boundary layer, the convective boundary layer and the stable boundary layer. We shall consider both the structure and the dynamics of these boundary-layer prototypes. Our discussion is limited to the so-called dry boundary layer, i.e. we shall disregard the influence of condensation and evaporation on boundary-layer dynamics.

Until now we have limited ourselves to a so-called horizontally homogeneous boundary layer. This means that all horizontal variability is neglected and all meteorological variables are thus only a function of the vertical distance above the surface. It will be clear that this allows considerable simplification in the description of the boundary layer. However, in practical circumstances horizontal variability can frequently not be omitted. Therefore, we shall review in the last part of this article two specific examples in which we study the influence of horizontal inhomogeneity on the structure of the atmospheric boundary layer. For the first example we consider the effect of a discontinuous change in surface boundary conditions, either as a change in roughness length or as a change in the surface heat flux. In the second example we treat the convective boundary both under influence of surface topography and of a horizontally varying heat flux.

Atmospheric turbulence determines not only the profiles of meteorological variables but it is also responsible for the dispersion of contaminants that are emitted in the boundary

A. Gyr and F-S. Rys (eds.), Diffusion and Transport of Pollutants in Atmospheric Mesoscale Flow Fields, 89–127.

layer. In other words, the concentration of air pollutants depends for a large part on the mixing properties of atmospheric turbulence. Therefore, we shall consider the dispersion characteristics for each of the boundary-layer prototypes mentioned above.

Finally, we should mention here that our knowledge on the atmospheric boundary layer has developed quite strongly over the last decades. In this review we can only present a limited selection of results that are now accepted as facts in the field of atmospheric boundary-layer meteorology. However, several textbooks and reviews have recently appeared to which a reader may turn for additional information. These are the reviews by Wyngaard (1988) and (1992). For an general introduction to atmospheric boundary-layer physics we can recommend the textbooks by Stull (1988) and Arya (1988). An more advanced textbook by Garratt (1992) gives an excellent survey of the state-of-the-art in modern atmospheric boundary-layer theory and its applications. For more information on atmospheric dispersion we may refer to Nieuwstadt and van Dop (1982) and to Venkatram and Wyngaard (1988).

4.2 Data

Clearly, any theory of the atmospheric boundary layer can only be accepted if it agrees with data. These are in the first place data gathered during measuring campaigns. Several of these field experiments in the boundary layer have been carried out recently and these have resulted in a wealth of observational data. A comprehensive list of these experiments with the type of measurements taken can be found in Stull (1989, pp 418-419). We shall use the data from several of these experiments in our discussion of the following sections.

However, an alternative to atmospheric experiments is presently available. This is numerical simulation. The supercomputers of today are so powerful that they allow us to solve the basic equations for a turbulent flow numerically. However, a computation of a turbulent flow in all its details, which is called direct numerical solution (DNS) or full numerical solution (FRS), is only possible under very restrictive assumptions. The reason is that any turbulent flow consists of a range of flow scales which extends from the largest scales, known as the macro structure, to the smallest scales, known as the micro structure. For the physics behind the existence of this macro and micro structure, we refer to a text book on turbulence, e.g. Tennekes and Lumley (1982). It turns out that the ratio of the smallest to the largest length scale in a turbulent flow is proportional $Re^{3/4}$. Here Re is the Reynolds number defined as

$$Re = \frac{u\ell}{\nu}$$

where u is a characteristic turbulence velocity scale, ℓ a representative macro length scale and ν the kinematic viscosity. It will be clear that for a FRS we require at least $Re^{9/4}$ grid points in our numerical model. Present computers can handle about 10^6 grid point and this means that the Reynolds number for our FRS is limited to $Re \simeq 500$. In the atmosphere, however, the values of Re can easily reach $5\,10^6$. Consequently, a FRS of the atmospheric boundary layer is out of the question.

Nevertheless, numerical simulation of atmospheric turbulence remains feasible if we restrict ourselves to the largest turbulence scales, i.e. the macro structure. This is known as large-eddy simulation (LES). The macro structure is isolated by applying a spatial filter to the complete turbulent flow field. This filter removes all fluctuations larger than a so-called filter length. Application of the filter to the equations that govern our turbulent flow, results in a set of equations for the filtered variables. This set must be solved numerically. However, the severe restriction on the numerical resolution is now relaxed. Namely, we can choose our filter length so that we can resolve the filtered turbulence on a numerical grid that our computer is still able to handle. However, we must remark here that the small flow scales that have been removed from our problem, cannot be completely neglected. They give rise to additional terms in the equations of the filtered variables. These terms require parameterization and this is called subgrid modelling. For further details we refer to the literature to be cited below.

One may wonder whether a filtered turbulent flow field still contains useful information. Here we should remember that the filtered turbulence represents the macro structure. It is a fact that this macro structure is characteristic for a particular turbulence flow and that it determines the turbulent transport properties. For instance, the exchange coefficient that relates the vertical fluxes to the gradients of mean variables, can be estimated as

$$K \simeq u\ell. \tag{4.1}$$

We will find that these parameters determine the vertical profiles of meteorological variables in our atmospheric boundary layer. Therefore, the data obtained by LES are quite useful for comparison with the theories and scaling relationships that we shall discuss in the following sections.

It perhaps not surprising that LES is becoming increasingly popular as a tool to study atmospheric turbulence. Nevertheless, its success varies with the type of boundary layer to which it is applied. LES has been the most successful for the convective boundary layer which, as we shall see in section 4.4.3, is dominated be large scale structures. Based on the results of a comparison study (Nieuwstadt et al. 1992), it was concluded that LES leads to a realistic simulation of this type of boundary layer. Moreover, it was found that a number of existing large-eddy codes performed about equally well and that the remaining difference between the codes could be only attributed to the subgrid parameterization.

This comparison exercise has been repeated for the neutral boundary layer (Anders et al. 1994). The various codes used led again to consistent and comparable results. However, the differences, which again could be mainly attributed to the subgrid model, were larger than in the comparison study for the convective boundary layer. LES of the stable boundary layer have only been performed quite recently (Mason and Derbyshire, 1990) so that it is to early to give a general comment on this application.

Here, we have only presented a limited and somewhat qualitative introduction to LES and its application to atmospheric turbulence. We must refer to the literature for a more extensive treatment. A general review of LES is given by Schumann and Friedrich (1989) and by Reynolds (1992). For a review on applications in the atmospheric boundary layer we refer to Schumann (1991). In the following sections we shall come back to the results of LES for our prototypes of the atmospheric boundary layer.

4.3 Scaling

In this section we consider the parameters and processes that determine atmospheric turbulence. Based on this we can define dimensionless quantities that allow us to introduce scaling regimes. Such scaling regime can be interpreted as an area in parameter space where boundary-layer dynamics is usually dominated by a single physical process. As a result, we can omit some of dimensionless parameters in a description of the boundary-layer that is valid in such scaling regime. This usually permits us to derive simplified expressions for the various meteorological variables. This is the approach that we shall take in the following sections.

First we consider the processes that produce turbulence in the boundary layer. Let us turn to the principal cause for existence of a boundary layer. It is the fact that the velocity must obey the no-slip condition at the surface. As a result the velocity must vary as a function of height. This velocity gradient is almost always unstable under the conditions that exist in the atmosphere and consequently turbulence is produced. This production process of turbulence is generally known as shear production.

The consequence of the no-slip condition is that the atmosphere experiences a shear force at the surface. This shear stress is denoted as τ_o. Let us use it to define a velocity scale

$$u_* = \sqrt{\frac{\tau_o}{\rho}} \tag{4.2}$$

where ρ is the density of the air. This velocity scale is known as the friction velocity. It (or τ_o itself) can be used the characterise turbulence due to shear production in the boundary layer.

Another process that influences turbulence in the atmospheric boundary layer, is related to density or temperature effects. The cause is the so-called sensible heat flux imposed on the boundary layer at the surface. The background of this heat flux is solar heating during daytime or radiative cooling during nighttime. Let us first consider a positive heat flux, i.e. heat is introduced in the boundary layer at the surface. This heat flux will cause an increase in temperature near the surface or alternatively a decrease in density. Due to the acceleration of gravity this less dense air will tend to rise. For the conditions found in the atmosphere, this process is again unstable and turbulence is generated. This production process is denoted as buoyant production.

However, the surface heat flux can be also negative, i.e. heat is removed from the boundary layer. The air near the surface will then be cooled and consequently its density increases. This situation is characterized as stable because the heavy air will resist vertical motion. As a result turbulence motion is opposed. We thus see the that production can be both positive and negative. Depending on circumstances, it can both produce or destroy turbulence. In this respect buoyant production is quite different from shear production.

The surface heat flux H_o that plays the dominant role in buoyant production, can be rewritten as

$$\overline{w'\theta'}_o = \frac{H_o}{\rho c_p}$$

where c_p is the specific heat of air at constant pressure. The $\overline{w'\theta}_o$ is known as the temperature flux and it consists of average correlation between vertical velocity and temperature fluctuations (the prime denotes a turbulent fluctuation and the overbar an appropriate mean). It shows that turbulent motions are responsible for the heat flux. We shall use $\overline{w'\theta'}_o$ to characterize turbulence due to buoyant production.

Based on the parameters u_* and $\overline{w'\theta'}_o$ introduced above we can define a length scale L given by

$$L = -\frac{u_*^3}{k\frac{g}{T_o}\overline{w'\theta'}_o} \tag{4.3}$$

where g (m/s^2) is the acceleration of gravity, T_o (K) the absolute temperature of the boundary layer (the combination g/T_o is sometimes denoted as the buoyancy parameter; it describes the vertical acceleration experienced by a temperature fluctuation of 1 K). The k is constant which is known as the Von-Karman constant. There are several values of this Von-Karman constant in use but at this moment 0.4 is the most accepted (Garratt, 1992; p 289). The L

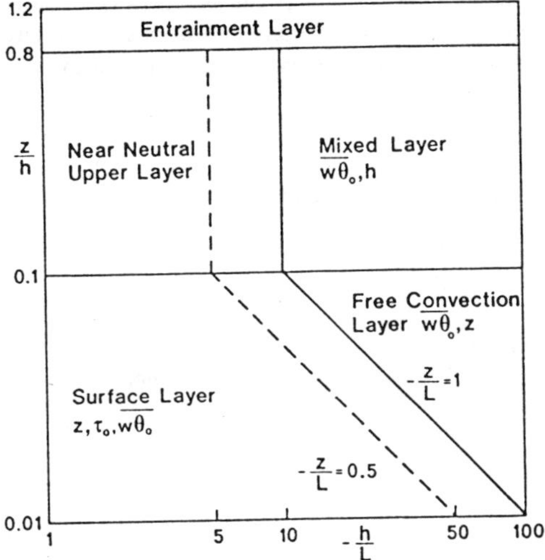

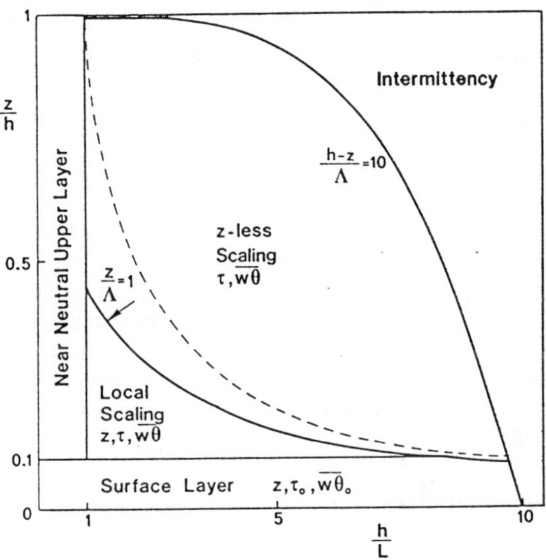

Figure 4.1: Scaling regions in the unstable ($\overline{w'\theta'}_o > 0$) and stable ($\overline{w'\theta'}_o < 0$) boundary layer; the lines denote the boundaries between the various scaling regimes (Holtslag and Nieuwstadt, 1986)

is known as the Obukhov length scale. It can be interpreted as the height above the surface where buoyancy production starts to dominate over shear production. This interpretation will become clear in section 4.4.1 when we introduce the ratio z/L where z is the height above the surface.

The introduction of a length scale as a parameter to characterize the relative importance of the two production processes of atmospheric turbulence may suggest a search for other length scales. It is known that the macro structure of turbulence scales with the geometry in which turbulence occurs. In our case this geometry is given by the boundary-layer height

$$h$$

which is usually defined as the depth of the turbulent layer above the surface.

Finally, another length scale that must play a role, is the height

$$z$$

above the surface.

With the three length scales introduced above we can define two dimensionless combinations. These are

$$\frac{z}{L} \quad \text{and} \quad \frac{h}{L}.$$

The parameter h/L is denoted as a stability parameter because it can be interpreted in terms of the relative importance of buoyancy production with respect to shear production. A values of $h/L < 0$ implies $\overline{w'\theta'_o} > 0$ so that buoyancy produces turbulence. For the case that this production process dominates (i.e. $-h/L \gg 1$) we call the boundary layer convective. When $h/L > 0$ or $\overline{w'\theta'}_o < 0$ buoyancy destroys turbulence and we call the boundary layer stable. For a value of h/L around zero density effects will have small influence on the structure and dynamics of the boundary layer. This case is identified as a neutral boundary layer.

The parameter z/h gives the location in the boundary layer. The region near the surface, i.e. $z/h \leq 0.1$, is usually denoted as the surface layer.

We can use both parameters h/L and z/h to define scaling regimes in the boundary layer. These scaling regimes are illustrated in Fig. 4.1. They can be interpreted as regions where boundary-layer turbulence can be typified in terms of a limited number of parameters. The selection of the these parameters follows from an identification of the physical processes that play a dominant role in each region. As the number of scaling parameters is limited, one is usually able to find simplified expressions to describe the boundary layer. In the following sections we shall discuss the background for this parameter selection and show how in each region we can formulate expressions for the boundary layer structure.

4.4 Homogeneous boundary layer

In this section we limit ourselves to the horizontally homogeneous boundary layer, i.e. a boundary layer without any horizontal variation so that profiles are only a function of height. Our treatment will be organized according to the prototypes of the boundary layer that we have mentioned in the introduction. For each case we first consider the characteristic scaling parameters and illustrate how we can use them to formulate expressions for various boundary-layer variables. Next we consider the dynamics of each boundary-layer prototype and we end with a discussion of its dispersion characteristics.

For a discussion of boundary-layer dynamics we must take into account the various time scales that can influence processes in the boundary layer. First we distinguish the so-called forcing time scale

$$\mathcal{T}_f.$$

It gives the time scale of the external processes that act upon the boundary layer. These are for instance the rotation of the earth expressed by the Coriolis parameter f (see Garratt, 1992 p. 24). We will take for this parameter a value representative for the middle latitudes, i.e. $f \simeq 10^{-4}$ s^{-1}. The forcing time scale $\mathcal{T}_f$ can also be taken as representative for the time variation of the surface boundary conditions such as the surface heat flux.

The second time scale is

$$\mathcal{T}_m$$

which describes the time development of the mean boundary-layer structure, e.g. the mean velocity or temperature profile.

Finally, we must consider the time scale of the turbulent transport processes in the boundary layer. It is given by

$$\mathcal{T}_t \simeq \frac{\ell}{u}$$

where ℓ and u are again a representative length and velocity scale for the turbulent macro structure.

At this stage, we can already make a simplification. For all the cases that we shall consider in this section on the homogeneous boundary layer we have

$$\mathcal{T}_t < \mathcal{T}_p$$

where $\mathcal{T}_p$ is the time scale on which the turbulent production processes, such as shear or buoyancy production, vary. This inequality means that at each time turbulence is in equilibrium with its production processes and e.g. memory effects can be neglected. The dynamics of turbulence is thus at all times determined by a balance between production and destruction processes. Such a condition is known as quasi-stationary turbulence.

4.4.1 Surface layer

Scaling

The surface layer is the region of the boundary layer close to the surface, i.e. $z/h < 0.1$. We assume that the length scale h does not play a role here. The background for this assumption is that h was taken as representative for the length scale of turbulence. However, near the surface the turbulent length scale will be rather determined by the distance to the surface than by the boundary-layer height. Moreover, we assume in the surface layer that

$$\frac{z}{z_0} \gg 1$$

where z_0 is a length that characterizes the roughness of the surface. This condition means that z_0 cannot influence directly the structure of the surface layer.

As a result of our discussion we find that only

$$u_* \quad L \quad \text{and} \quad z$$

remain as scaling parameters in the surface layer where u_* and L have been already defined in (4.2) and (4.3) respectively. Our hypothesis is that these parameters form a complete set,

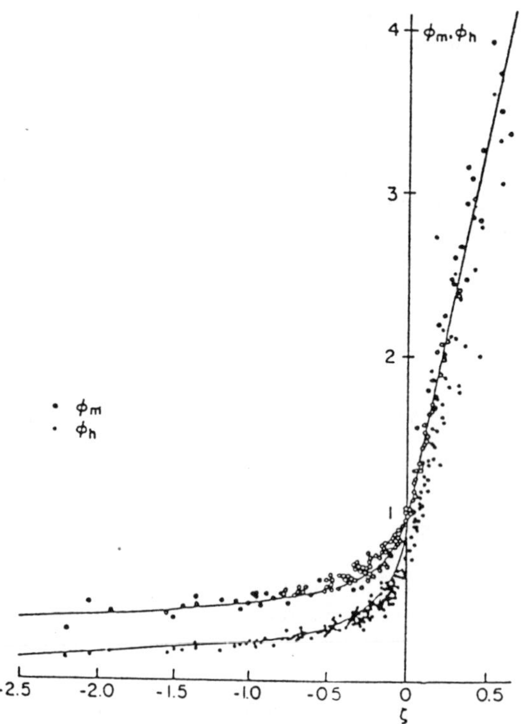

Figure 4.2: The flux relationships ϕ_m and ϕ_h in the surface layer scaled according to Monin-Obukhov similarity where ζ stands for z/L; the lines denote empirical expressions such as discussed in the text and the symbols are observations (Fleagle and Businger, 1980)

i.e. all turbulent variables scales in terms of these parameters follow a unique relationship as function of z/L. This is called Monin-Obukhov similarity. The ratio z/L can be interpreted as describing the influence of stability effects in the surface layer. When $z/L \approx 0$ stability effects can be neglected. We see that this occurs for small values of z or for small values of $\overline{w'\theta'}_o$ with respect to u_*. In other words stability effects have only effect at sufficient large values of z.

Let us take as example of Monin-Obukhov similarity the so-called flux relationships. These are the dimensionless velocity and temperature gradient defined by

$$\begin{aligned} \phi_m\left(\frac{z}{L}\right) &= \frac{kz}{u_*}\frac{\partial U}{\partial z} \\ \phi_h\left(\frac{z}{L}\right) &= \frac{kz}{\theta_*}\frac{\partial \Theta}{\partial z} \end{aligned} \tag{4.4}$$

where U and Θ are the mean velocity and temperature in the surface layer and where the temperature scale θ_* is defined as $-\overline{w'\theta'}_o/u_*$. In Fig. 4.2 we show a large number observations of. The fact that the observations plot on a single curve is an indirect proof for the validity of this scaling approach. From this figure it becomes also clear that the value for the Von-

Karman constant, k, has been chosen such that $\phi_m(0) = \phi_h(0) = 1$.

Based on data shown in Fig. 4.2 we can already learn some facts about the influence of stability on turbulence. For $z/L < 0$, which implies unstable or convective conditions, we have argued that turbulence is produced by buoyancy. We may expect in that case a large turbulence intensity and thus strong vertical transport. This will result in small gradients of the mean profiles and consequently in small values of the gradient functions ϕ_m and ϕ_h. The opposite is the case for stable condition when $z/L > 0$. Turbulence is then reduced by buoyancy effects and vertical transport becomes small. As a result the mean gradients and consequently the values of ϕ_m and ϕ_h will be large.

The expressions for ϕ_m and ϕ_h allow also another interpretation. Let us assume that the turbulent fluxes of momentum, τ, and of temperature, $\overline{w'\theta'}$, are approximately constant throughout the surface layer. This means that they are equal to their values at the surface that we have defined in section 4.3 as u_*^2 and $\overline{w'\theta'}_o$. The expressions (4.4) can now be rewritten as

$$\begin{aligned} \tau &= K_m \frac{\partial U}{\partial z} \\ \overline{w'\theta'} &= K_h \frac{\partial \Theta}{\partial z} \end{aligned}$$

with

$$K_m = \frac{k u_* z}{\phi_m(z/L)} \quad \text{and} \quad K_h = \frac{k u_* z}{\phi_h(z/L)}. \tag{4.5}$$

In other words the turbulent fluxes in the surface layer can be parameterized in terms of so-called K-theory. We shall return to the consequences of this result in the next sections

The functions ϕ_m and ϕ_h are still unknown. They must follow from fitting expressions to the data given in Fig. 4.2. Several of such expressions have been proposed in the literature (see e.g. Sorbjan, 1988; p. 74-76). An example is

$$\begin{aligned} &\frac{z}{L} < 0 \qquad \phi_m = (1 - 16\frac{z}{L})^{-1/4} \quad \phi_h = (1 - 14\frac{z}{L})^{-1/2} \\ &\frac{z}{L} > 0 \qquad \phi_m = 1 + 5.2\frac{z}{L} \qquad\quad \phi_h = 1 + 5.2\frac{z}{L}. \end{aligned} \tag{4.6}$$

Monin-Obukhov similarity can be also applied to other variables in the surface layer. Let us consider σ_w and σ_θ which are the standard deviations of the verical velocity fluctatuations and the temperature fluctuations, respectively. According to Monin-Obukhov similarity it should follow that

$$\begin{aligned} \frac{\sigma_w}{u_*} \equiv \frac{\sqrt{\overline{w^2}}}{u_*} &= f_w\left(\frac{z}{L}\right) \\ \frac{\sigma_\theta}{\theta_*} \equiv \frac{\sqrt{\overline{\theta^2}}}{\theta_*} &= f_\theta\left(\frac{z}{L}\right). \end{aligned} \tag{4.7}$$

In Fig. 4.3 we have plotted some observations of both σ_w and σ_θ in the surface layer. The data plot indeed on a single curve which is again a confirmation of Monin-Obukhov similarity.

It may seem that Monin-Obukhov similarity is generally valid in the surface layer. However, for some parameters Monin-Obukhov similarity fails. An example is the standard deviation of the horizontal velocity fluctuations σ_u and σ_v. The explanation of this failure will be deferred to section 4.4.3

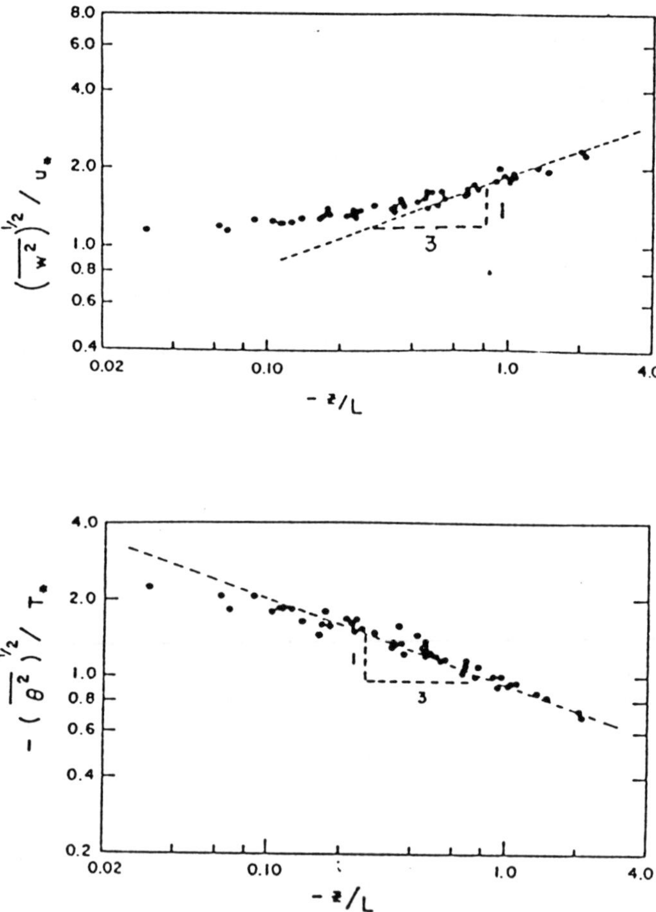

Figure 4.3: The standard deviation of the vertical velocity fluctuations (top) and of the temperature fluctuations (bottom) as function of z/L according to Monin-Obukhov similarity; the dashed line is discussed in the section on the convective boundary layer; the symbols denote experiments (Wyngaard et al. 1971)

Dynamics

We have shown in the previous section that tranport processes in the boundary layer can be parameterized in terms of K-theory. This means the time-evolution equation for the mean velocity U and temperature Θ can be written as

$$\begin{aligned} \frac{\partial U}{\partial t} &= \frac{\partial}{\partial z}\left(K_m \frac{\partial U}{\partial z}\right) \\ \frac{\partial \Theta}{\partial t} &= \frac{\partial}{\partial z}\left(K_h \frac{\partial \Theta}{\partial z}\right). \end{aligned} \tag{4.8}$$

Based on these equation it follows that the time scale $\mathcal{T}_m$ for these surface-layer variables at height z can be estmated as

$$\mathcal{T}_m \simeq \frac{z^2}{K} \simeq \frac{z}{u_*} \tag{4.9}$$

where in the second step we have used (4.5) with the appoximation ϕ_m and $\phi_h \simeq 1$.

However, the ratio z/u_* is also representative for the turbulence time scale in the surface layer because the turbulence length scale $\ell \simeq z$ and the velocity scale $u \simeq u_*$. In other words for the surface we find

$$\mathcal{T}_m \simeq \mathcal{T}_t.$$

The consequence of these estimates is that the time scale $\mathcal{T}_m$ is in general small with respect to the forcing time scale $\mathcal{T}_f$. Let us take for $\mathcal{T}_f$ the time scale of rotation, i.e. the inverse of the Coriolis parameter f^{-1}. We then have $\mathcal{T}_f \approx 3$ hours whereas with $z \simeq 10$m and $u_* \sim 0.3$ m/s $\mathcal{T}_m \approx 30$ s.

As a result of $\mathcal{T}_m \ll \mathcal{T}_f$, the dynamics of the mean profiles in the surface layer can be characterized as quasi-stationarity and we will expect no influence of rotation. Quasi-stationarity means here that the time derivative in (4.8) can be neglected. This is at the same time the justification of our assumption made the previous section that turbulent fluxes in the surface layer are independent of height.

The equations (4.6) can be integrated. The result is the vertical profile for the mean velocity and temperature in the surface layer. Because of quasi-stationarity, these profiles depend only indirectly on time through the time varying values of u_* and $\overline{w'\theta'}_o$. For the closed-form expressions of U and V we refer to Garratt (1992, p. 52). For the case of a neutral surface layer (i.e. $z/L = 0$) we find the well-known logarithmic profile

$$U = \frac{u_*}{k} \ln\left(\frac{z}{z_0}\right) \tag{4.10}$$

Dispersion

We have seen above that vertical transport processes in the surface layer can be parameterized in terms of K-theory. This means also that the vertical dispersion of a contaminant in the surface layer can be described by a diffusion equation.

Let us consider the dispersion of the mean concentration C emitted from an instantaneous point source with strength Q. The emission height z_s is at ground level ($z_s = 0$). The diffusion equation for this problem reads

$$\frac{\partial C}{\partial t} = \frac{\partial}{\partial z}\left(K_h(z)\frac{\partial C}{\partial z}\right). \tag{4.11}$$

We have set the scalar diffusion coefficient equal to the diffusion coefficient for heat (4.5).

For the case of the neutral surface layer, when $K_h = ku_*z$, the solution of this problem becomes

$$C = \frac{Q}{\overline{z}} e^{-z/\overline{z}}. \tag{4.12}$$

where the height, $\overline{z}$, of the concentration distribution is defined as

$$\overline{z} = \frac{1}{Q} \int_0^\infty zC(z)\, dz = ku_*t. \tag{4.13}$$

The ground-level concentration, $C_o = C(0)$, thus decreases as t^{-1}.

Next we consider a continuous point source with strength Q which is described by the diffusion equation

$$U(z)\frac{\partial C}{\partial x} = \frac{\partial}{\partial z}\left(K_h(z)\frac{\partial C}{\partial z}\right). \tag{4.14}$$

where for the case of a neutral surface layer $U(z)$ is given by (4.10).

For the case $z_s = 0$, no exact solution is known for this problem. However, based on some approximations (Hunt, 1982) we may find

$$\overline{z} \simeq x^{0.9}.$$

We see that the $\overline{z}$ increases slower than in the case of the instantaneous point source. This is caused by fact that the wind velocity in (4.14) increases as a function of height so that concentration at high levels will be transported downstream faster than concentration at a low heights. As a result the growth rate of plume height is decreased.

Until now we have restricted ourselves to a ground-level source, i.e. $z_s = 0$. For the case of an elevated source, i.e. $z_s > 0$, we must consider the size of the plume. For this we take the dispersion coefficient σ_z which defined as

$$\sigma_z^2 = \overline{(z - z_s)^2} = \frac{1}{Q}\int_{-\infty}^{\infty} (z - z_s)^2 C(z)\, dz. \tag{4.15}$$

Initially, σ_z is smaller than the turbulence length scale $\ell \sim z$. In that case dispersion can not be described be a standard diffusion equation such as (4.11) or (4.14). Instead we should apply Taylors' expression which for the instantaneous source leads to

$$\sigma_z^2 = \sigma_w(z_s)t^2.$$

Only for $\sigma_z \geq z$, i.e when the plume has reached the surface, we can apply again the diffusion equation. For further details we refer to Hunt (1982).

4.4.2 Neutral boundary layer

Scaling

In the following section we consider the region of the boundary layer above the surface layer, i.e. $z/h > 0.1$. For the case of the neutral boundary layer $\overline{w'\theta'}_o = 0$ or buoyancy effects do not play a role. As scaling parameters then remain

$$u_* \quad h \quad \text{and} \quad z$$

Before we consider scaling relationships for the neutral boundary layer, one may ask whether this type of boundary layer can occur in practice. For this we have to estimate the size of temperature fluctuations above which we can no longer neglect buoyancy. For this we introduce

$$Ri' = \frac{g}{T_o}\frac{\theta' h}{u_*^2}$$

which gives the ratio of the acceleration due to a temperature fluctuation θ' with respect to the turbulent acceleration u_*^2/h. We assume that for $Ri' < 0.1$ buoyancy effects can be neglected. If we take $h = 500$ m, $g/T_o = 0.03$ m°C/s and $u_* = 0.3$ m/s, we find that this condition is only satisfied for $\theta' < 6\,10^{-4}$ °C. We must therefore conclude that even small temperature fluctuations quickly lead to appearance of buoyancy effects in the boundary layer. Small temperature fluctuations are almost always present so that the case of the neutral boundary layer must be considered as an exception which will be rarely encountered in the real atmosphere.

Nevertheless a study of the neutral boundary layer is useful because it allows us isolate the influence of shear production on turbulence. However, it will be clear that observations will be difficult to come by. An exception are the experimental data described by Grant (1992). Most information on this type of boundary layer thus follows from large-eddy simulations. An example is the study by Mason and Thomson (1987). For an investigation in which various large-eddy codes for the neutral boundary layer have been compared, we refer to Andren et al. (1994).

Dynamics

For a shear dominated boundary layer the turbulent length scale ℓ can be estimated as $0.07h$ (Hinze, 1975). This means that the turbulence scale is much smaller than the boundary-layer depth h which is the distance over which mean variables such as U change.

The condition $\ell \ll h$ is a necessary condition for the applicability of K-theory. It seems thus reasonable to estimate vertical transport in a neutral boundary layer in terms of K-theory. Following a similar reasoning as used for the surface layer we then may derive for the time scale of the mean boundary layer structure

$$\mathcal{T}_m \simeq \frac{h^2}{K} \simeq \frac{14h}{u_*}$$

where in the second step we have used (4.1) and the estimate for ℓ given above. If we take as representative values $h \simeq 500$ m and $u_* \simeq 0.3$ m/s, $\mathcal{T}_m$ becomes of the same order of magnitude as the $\mathcal{T}_f$ which is estimated as $f^{-1} = 10^4$ s.

We may thus conclude that in contrast to the surface layer the mean structure of the neutral boundary layer can not be considered as quasi-stationary and the influence of Coriolis effects can not be neglected. In other words, to describe the neutral boundary layer one has to solve the full time-dependent equations such as (4.8). (Note that in (4.8) the influence of rotation should still be added).

Dispersion

Dispersion processes in the neutral boundary layer may be treated along the same line as we have applied in the surface layer. That means that for a surface source (i.e. $z_s = 0$) we may use a diffusion equation. However, the diffusion coefficient, K_h, can no longer be set equal to ku_*z. We should take into account that that the turbulence length scale ℓ is limited to $0.07h$. This puts a limit on the value of K_h. A rough estimate for the maximum diffusion coefficient

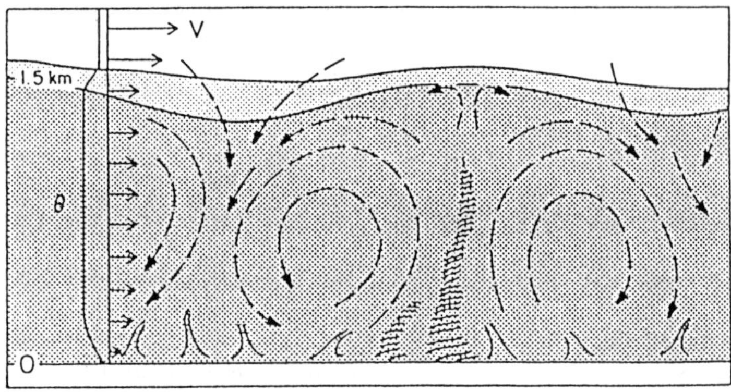

Figure 4.4: Schematic picture of the convective boundary layer (Wyngaard, 1992)

in the neutral boundary layer is $0.07u_*h$. Near the top of the boundary layer, we may expect that the diffusion coefficient should approach zero. An approximate interpolation equation between these limits is

$$K_h = ku_*z(1-\frac{z}{h})^2.$$

For a non-surface source we must again refrain from using the diffusion equation before the concentration distribution has reached the surface, i.e. before $\sigma_z > \ell$. For $\sigma_z < \ell$, we may again use Taylors' expression as a first approximation.

4.4.3 Convective boundary layer

Scaling

In fig. 4.4 we show a schematic picture of the convective boundary layer, i.e. $\overline{w'\theta'}_o > 0$. We see that the turbulence structure is composed of large eddies which fill the whole boundary layer. These eddies are a direct result of buoyancy production. They consist mainly of air heated up at the surface. Due to buoyancy, this air will tend to rise. Consequently these eddies are known as thermals or plumes. The influence of shear production can be neglected and this means that u_* plays no longer a role. Based on this consideration we propose as scaling parameters for the convective boundary layer

$$h \quad w_* \quad \text{and} \quad z$$

where the so-called convective velocity scale is defined as

$$w_* = \left(\frac{g}{T_o}\overline{w'\theta'}_o h\right)^{\frac{1}{3}}. \tag{4.16}$$

The scaling of variables in terms of the parameters given above is known as mixed-layer scaling. As example we shown in Fig. 4.5 the vertical profiles of σ_w, σ_u and the σ_θ. According to mixed-layer scaling these variables should obey the following expressions

$$\frac{\sigma_w}{w_*} \equiv \frac{\sqrt{\overline{w^2}}}{w_*} = f_w\left(\frac{z}{h}\right)$$

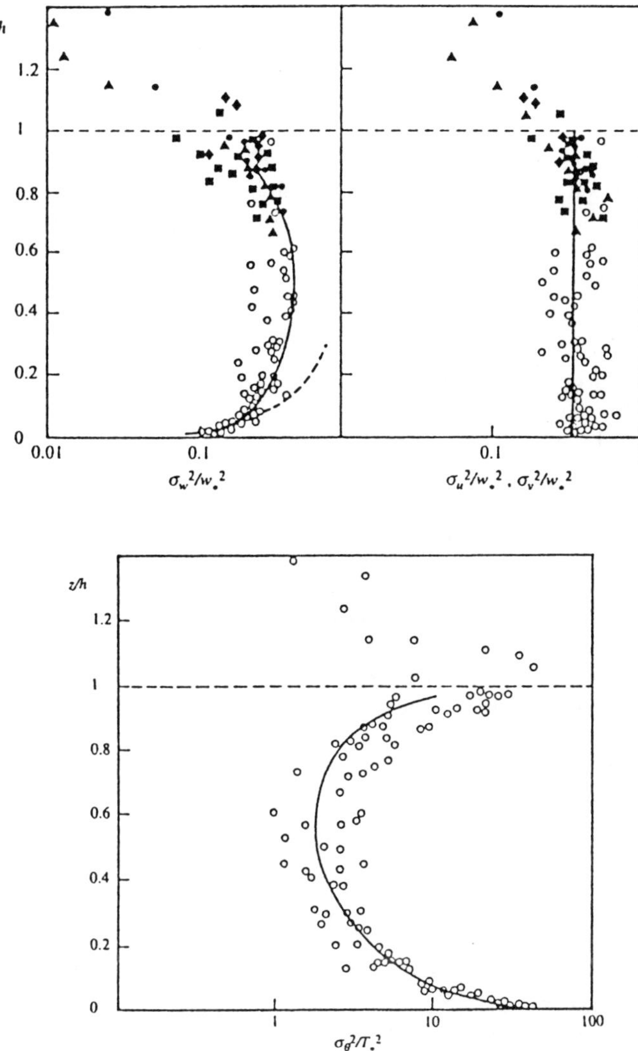

Figure 4.5: Vertical profiles of the standard deviation of vertical velocity fluctuation, horizontal velocity fluctuations (top) and temperature fluctuations (bottom) scaled following mixed layer scaling; the lines denote empirical fits to the observations indicated by the symbols (Garratt, 1992)

$$\begin{aligned} \frac{\sigma_u}{w_*} \equiv \frac{\sqrt{\overline{u^2}}}{w_*} &= f_u\left(\frac{z}{h}\right) \\ \frac{\sigma_\theta}{T_*} \equiv \frac{\sqrt{\overline{\theta^2}}}{T*} &= f_\theta\left(\frac{z}{h}\right) \end{aligned} \tag{4.17}$$

where the temperature scale T_* is defined as $T_* = \overline{w'\theta'}_o/w_*$. We find that the data in Fig. 4.5 plot reasonably well on a single curve which confirms mixed-layer scaling.

One may wonder how the expressions (4.17) which are valid in the mixed-layer, should be matched to the expressions (4.7) that we have derived for the surface layer. This requires a matching condition which can be expressed as

$$\lim_{z/h\to 0} (\text{mixed layer}) = \lim_{-z/L\to\infty} (\text{surface layer})$$

This condition can be alternatively expressed as: in the matching region between the mixed-layer and the surface layer only $\overline{w'\theta'}_o$ and z remain as scaling variables. The region where this is valid, is called the free convection layer (see Fig 4.1).

If we apply the matching condition to the expressions f_w and f_θ given by (4.7) and (4.17) we find

$$\begin{aligned} \frac{\sigma_w}{u_*} &\simeq 1.8\left(-\frac{z}{L}\right)^{1/3}, \\ \frac{\sigma_\theta}{\theta_*} &\simeq 0.9\left(-\frac{z}{L}\right)^{-1/3} \end{aligned}$$

where the coefficients are obtained by fitting to experimental data. The result of matching is an explicit relationship for these standard deviations as a function of height. This follows from the fact that in the free convection layer only a limited number of scaling parameters remain. If we return to Fig. 4.3, we see that the free-convection expressions are very well followed by the experimental data.

If we consider the horizontal velocity fluctuations given by (4.17) in the free convection layer, i.e. for $\lim_{z/h\to 0}$, we find

$$\sigma_u \simeq 0.6 w_* \simeq 0.8\left(-\frac{h}{L}\right)^{1/3} u_*$$

where we have used the data shown in Fig. 4.5. From this expression it clear that σ_u does not approach Monin-Obukhov similarity, as we already have mentioned in section 4.4.1. Here, we find the reason. Namely, near the surface h remains important as a length scale.

Mixed-layer similarity and free convection scaling is also confirmed by the results of large-eddy simulation. Several LES simulations on the convective boundary layer have appeared in the literature. Examples are Schmidt and Schumann (1988) and Mason(1989). The success of large-eddy simulation is due to that fact that the convective boundary layer is dominated be large-scale eddies that can be easily resolved numerically. For review that simultaneously consists of a comparison study of several large-eddy codes, we refer to Nieuwstadt et al. (1992).

The numerical simulations results allow us to study the turbulent structure of the convective boundary layer in great detail. For instance we are able to calculate conditional statistics that we can use to investigate the dynamics of the thermals (Schumann and Moeng, 1991). Moreover, large-eddy simulation can be used to study dispersion in the convective boundary of which we will see some examples below.

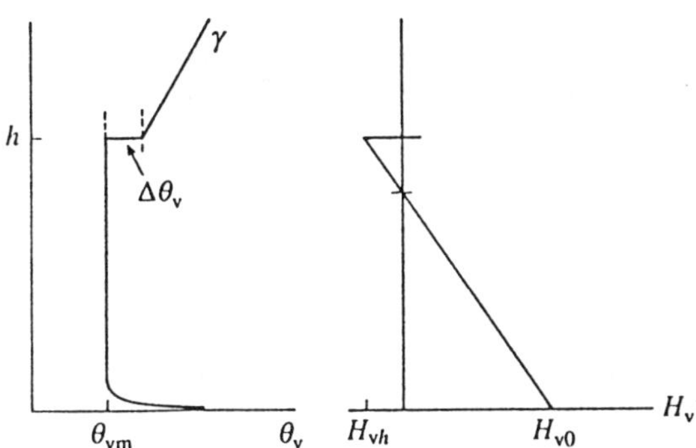

Figure 4.6: Schematic representation of the mean temperature (left) and the temperature flux (right) profile in a convective boundary layer (Garratt, 1992)

Dynamics

We have seen in Fig. 4.4 that turbulent eddies in the convective boundary layer fill the whole boundary layer so that the length scale of turbulence $\ell \simeq h$. In order words, the turbulence dynamics is dominated by large scale vertical transports and consequently K-theory fails. Moreover, these large scale transports imply that the time scale of the mean structure is close to the turbulent time scale: $\mathcal{T}_m \simeq \mathcal{T}_t$. The $\mathcal{T}_t$ may be estimated as h/w_*. With typical for $h \sim 1000$ m and $w_* = sim1$ m/s we find that $\mathcal{T}_t \sim 15$ min.

If we again take for the time scale of forcing $\mathcal{T}_f$ the rotation time scale f^{-1}), we find that

$$\mathcal{T}_m \ll \mathcal{T}_f.$$

This leads us to the conclusion that the mean structure of the convective boundary layer is quasi-stationary and also that the effects of rotation may be neglected.

Because the turbulence intensity is large, one may expect efficient vertical mixing in a convective boundary layer. As a result the mean variables can be described by a well-mixed profile as illustrated in Fig. 4.6 for the average temperature.

With this time-invariant shape of $\Theta(z)$ as point of departure, we consider the energy conservation equation which reads

$$\frac{\partial \Theta}{\partial t} = -\frac{\partial \overline{w\theta}}{\partial z}. \tag{4.18}$$

Let us integrate this equation between $0 < z < h$ and $h^- < z < h^+$ with as result

$$\begin{aligned} \frac{d\Theta}{dt} &= \frac{\overline{w\theta}_o - \overline{w\theta}_h}{h} \\ \Delta\Theta \frac{dh}{dt} &= -\overline{w\theta}_h \end{aligned} \tag{4.19}$$

where $\Delta\Theta = \Theta(h^+) - \Theta(h^-)$. For $\Delta\Theta$ we can derive the equation

$$\frac{d\Delta\Theta}{dt} = \gamma \frac{dh}{dt} - \frac{d\Theta}{dt}. \tag{4.20}$$

The first equation of (4.19) implies that the temperature flux is a linear function of z with a value of $\overline{w\theta_h}$ at the top of the boundary layer. Such linear profile is the consequence of (4.18) together with the assumption the shape of Θ is time invariant.

We find that the set of equations (4.19) and (4.20) contains more unknowns than equations. We have thus a closure problem. This is solved be expressing the $\overline{w\theta_h}$ in terms of surface temperature flux

$$\overline{w\theta_h} = -\beta\,\overline{w\theta_o} \tag{4.21}$$

with the constant β equal to $0.1 - 0.2$. For a more detailed discussion of this so-called entrainment assumption we refer to Garratt (1992, p. 158).

The set of equations (4.19), (4.20) and (4.21) can be solved explicitly. For the case of a constant $\overline{w\theta_o}$ we find

$$h^2 - h_0^2 = 2\frac{1+2\beta}{\gamma}\,\overline{w\theta_o}\,t \tag{4.22}$$

where h_0 is the initial height of the boundary layer at $t = 0$.

With this equation and with a given value for $\overline{w\theta_o}$ we can make an estimate of the boundary-layer height. A representative value for $\overline{w\theta_o}$ is 0.1 m°C/s. For γ we take 0.01 °C/m. We then find from (4.22) that over a period of 8 hours the convective boundary layer can grow by $\sim$ 900 m. The convective boundary layer is thus in general quite deep. This is consistent with the large turbulence activity and the strong mixing in this type of boundary layer.

With $h = h(t)$ all the parameters for mixed-layer scaling are known so that we can determine the structure of our boundary layer at each time instant.

Dispersion

At first we shall restrict ourselves to an instantaneous surface source, i.e. $z_s = 0$. The height of the concentration distribution can be described by $\overline{z}$ defined in (4.13). When this height is still much smaller than h, we can apply free convection scaling. These means that the only scaling parameters which determine the concentration distribution are the $g/T_o\overline{w'\theta'}_o$ and the dispersion time t so that

$$\overline{z} \sim \left(\frac{g}{T}\overline{w'\theta'}_o\right)^{1/2} t^{3/2}. \tag{4.23}$$

We have already seen in (4.12) that the surface concentration C_o can be directly related to $\overline{z}$ with as result

$$C_o \sim \frac{Q}{\sigma_z} \sim t^{-3/2}. \tag{4.24}$$

This expression which is valid for an instantaneous source can be extended to a continuous source by using the transformation $t = x/U$ where U is the mean velocity. We can apply this transformation, because in the well-mixed convective boundary layer U is independent of height. Substituting this transformation in (4.24) and using mixed-layer scaling we find

$$C_o = 0.9X^{-3/2} \tag{4.25}$$

where the similarity variable X is defined as

$$X = \frac{xw_*}{hU}$$

and where the factor 0.9 follows from a fit to experimental data. This relationship is illustrated in Fig 4.7 and we see that it agrees very well with the experimental data.

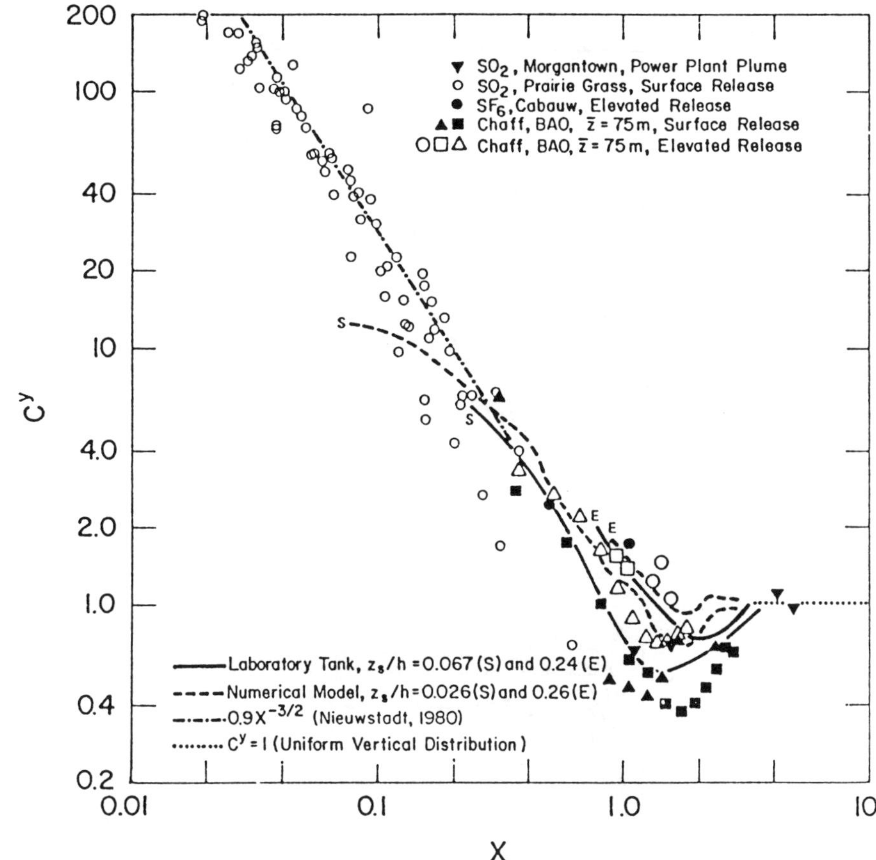

Figure 4.7: The non-dimensional ground-level concentration emitted from a surface source in the convective boundary layer as a function of non-dimensional dispersion distance $X = w_* x/(Uh)$; the lines denote various approximations where in particular the dashed-dotted line gives the expression for the free convection scaling discussed in the text; the symbols denote various experimental data (Briggs, 1992)

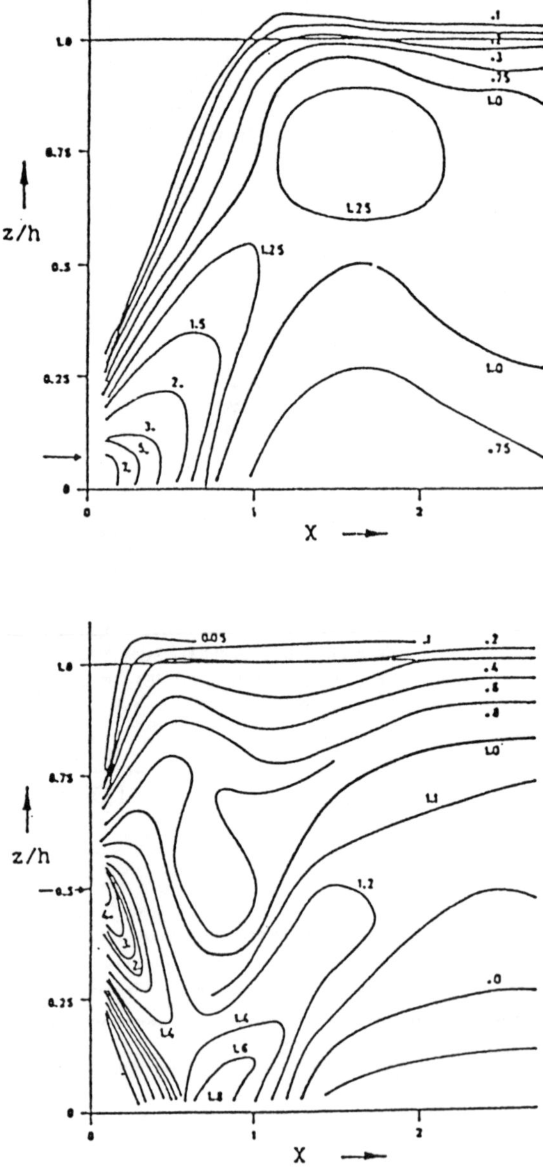

Figure 4.8: Vertical concentration distribution as a function of non-dimensional dispersion distance x for a low-level source (top) and an elevated source (bottom) as a function of the non-dimensional dispersion distance $X = w_* x/(Uh)$ (Willis and Deardorff, 1976 and 1982)

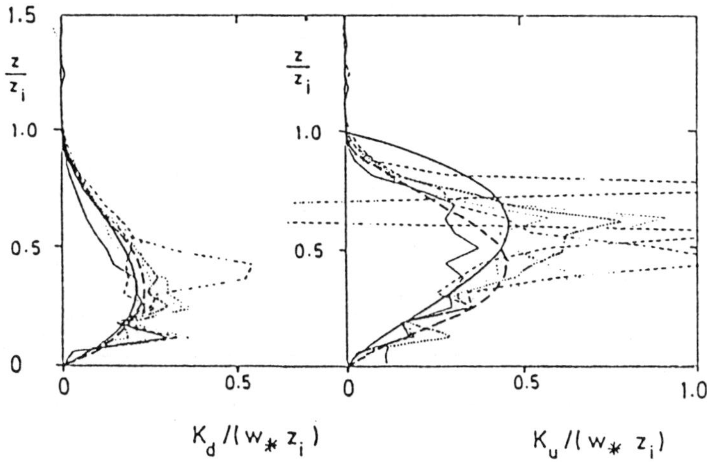

Figure 4.9: The diffusion coefficients for top-down (K_d) and for bottom-up dispersion (K_u) as calculated for several runs with a large-eddy simulation (Schumann, 1989)

A surprising characteristic of dispersion in the convective boundary layer for other source geometries is illustrated in Fig. 4.8. Here, we show the concentration pattern of a low level and an elevated source as a function of the diffusion distance denoted by the variable X defined above. The data were obtained in a laboratory experiment by Willis and Deardorff (1976, 1981). For the case of the low-level source we see that the plume rises strongly and even causes an elevated concentration maximum near the top of the boundary layer. The plume from the elevated source descends and impinges on the ground where it causes a local concentration maximum. It should be stressed that it is not possible to obtain such plume behaviour as a solution of a diffusion equation. This is another confirmation that K-theory is not applicable to describe transport processes in the convective boundary layer.

The reason for the anomalous behaviour shown in Fig 4.8 is the vertically inhomogeneous and asymmetric structure of turbulence in the convective boundary layer. For instance, according to Fig. 4.5 we find that the vertical velocity fluctuations increase strongly near the surface as a function of height. Such vertical inhomogeneity is the background of the plume lift-off illustrated in Fig. 4.8. The downward motion of the elevated plume can be related to the thermals in which, as mentioned above, the air heated at the surface is transported upward. Due to that fact that the upward velocity is concentrated in these isolated thermals we find that the distribution of the vertical velocity fluctuations is skew with a negative mode. This negative mode causes the plume axis to descend.

To elaborate on this strong asymmetry in plume dispersion, we consider some results obtained with a large-eddy simulation. We have already mentioned that this technique has been used with much success to study the turbulent structure of the convective boundary layer. Here, we apply it to the dispersion of continuous source in this boundary layer. The location of the source is either at ground level or at the top of the boundary layer. The first case will for obvious reasons be called bottom-up and the second top-down diffusion, a term coined by (Wyngaard and Brost, 1984). In Fig. 4.9 we illustrate the diffusion coeffi-

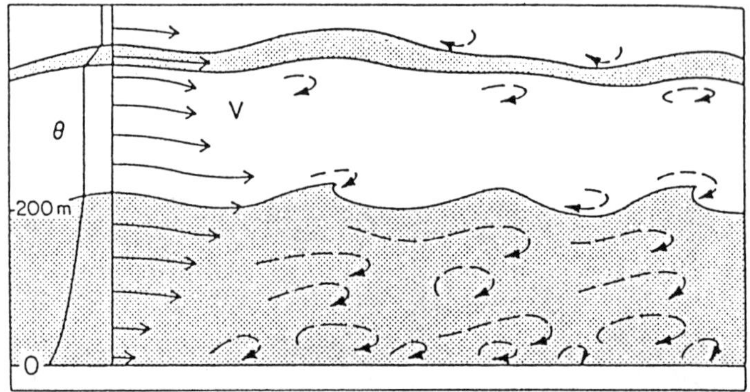

Figure 4.10: Schematic picture of the stable boundary layer (Wyngaard, 1992)

cients for both cases, which are calculated as the vertical concentration flux divided by the mean concentration gradient. We notice that the top-down diffusion coefficient (K_d) is much small than the bottom-up coefficient (K_u). The explanation lies again in the effect of the thermals. This implies the the K-coefficient is process dependent and as such the concept of K-theory, in which the K-coefficient is a unique function of position, fails. This is another clear confirmation that K-theory is not applicable to the convective boundary layer.

For a further discussion on dispersion in the convective boundary layer we refer to Lamb (1982), Briggs (1988) and Weil (1988).

4.4.4 Stable boundary layer

Scaling

A schematic picture of the stable boundary layer is shown in Fig. 4.10. Compared with the convective boundary layer illustrated in Fig. 4.4 we see that the structure of the stable boundary layer is completely different. Most noticeable is that the boundary layer is rather shallow. Moreover, we have argued that in a stable boundary layer buoyancy effects oppose vertical motion. Consequently, turbulent eddies will be small with as result that the length scale $\ell \ll h$.

Shear production remains as the only source of turbulence and it will be clear that turbulence must be scaled in terms of this process. However, the correct scaling velocity is not the surface friction velocity u_*. Because of the fact that $\ell \ll h$, turbulence at each height z will be only determined by local processes. In other words turbulent eddies do not feel the influence of the surface directly. As a result the appropriate scaling parameters are

$$\overline{w'\theta'} \quad \tau \quad \text{and} \quad z$$

where $\overline{w'\theta'}$ and τ stand for the turbulent temperature and momentum flux at height z. With these parameters and the buoyancy parameter g/T_o we can define a length scale

$$\Lambda = -\frac{(\tau/\rho)^{3/2}}{k\frac{g}{T_o}\overline{w\theta}} \tag{4.26}$$

which is known as the local Monin-Obukhov length.

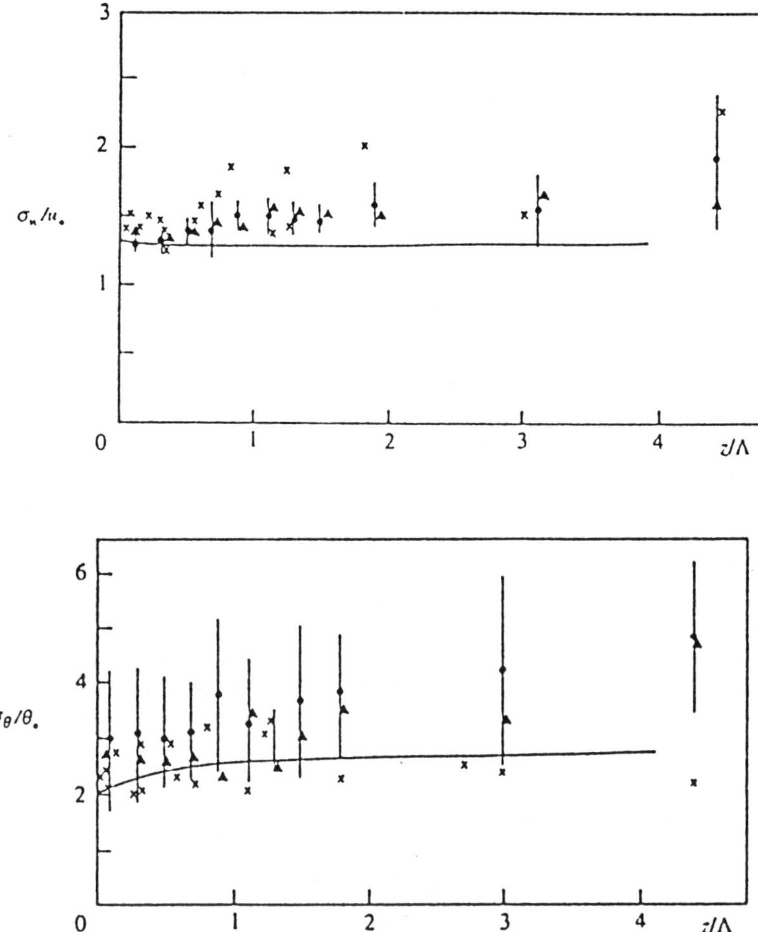

Figure 4.11: The standard deviation of the vertical velocity fluctuations (σ_w) (top) and the temperature fluctuations (σ_θ) (bottom) as function of z/Λ according to local scaling; the solid line is second-order turbulence model; each symbol denotes the average over several observations and the vertical lines given the variation among these observations (Nieuwstadt, 1984a).

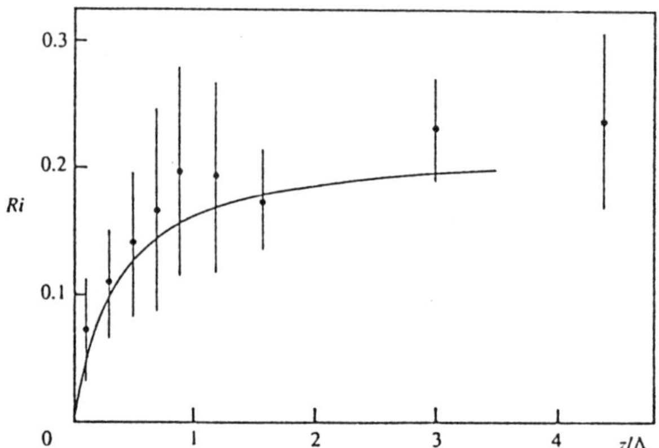

Figure 4.12: The Richardson number Ri as function of z/Λ according to local scaling; the solid line is second-order turbulence model; each symbol denotes the average over several observations and the vertical lines given the variation among these observations (Nieuwstadt, 1984a)

Scaling in terms of the parameters mentioned above is known as local scaling (Nieuwstadt, 1984a). We note that local scaling resembles the Monin-Obukhov similarity of the surface layer discussed in section 4.4.1 in which u_* and L are used as scaling parameters. As a consequence local scaling matches automatically to Monin-Obukhov similarity when $z \rightarrow 0$.

Application of local scaling to the standard deviation of the vertical velocity and of the temperature fluctuations leads to

$$\begin{aligned} \frac{\sigma_w}{\sqrt{\tau/\rho}} \equiv \sqrt{\frac{\overline{w^2}}{\tau/\rho}} &= f_w\left(\frac{z}{\Lambda}\right) \\ \frac{\sigma_\theta}{\overline{w\theta}/\sqrt{\tau/\rho}} \equiv \sqrt{\frac{\overline{\theta^2}}{\overline{w\theta}^2\tau/\rho}} &= f_\theta\left(\frac{z}{\Lambda}\right). \end{aligned} \tag{4.27}$$

These expression are compared with observations in Fig. 4.11. Taking into account the large scatter which is customary for measurements in the stable boundary layer, we find reasonable confirmation for our local scaling approach.

Another parameter to which we may apply local scaling is the so-called gradient Richardson number Ri defined as

$$Ri = \frac{\frac{g}{T_o}\frac{\partial \Theta}{\partial z}}{(\frac{\partial U}{\partial z})^2} \tag{4.28}$$

which quantifies local stability. According to local scaling Ri should be only a function of z/Λ. We find in Fig 4.12 that on the average the data of Ri plot reasonable well on a single curve.

We note that the curve in Fig. 4.12 approaches a constant value for $z/\Lambda \rightarrow \infty$. This is characteristic for another scaling regime in the stable boundary layer which is known as z-less scaling (see Fig 4.1). The background is that at large heights in the stable boundary layer, i.e. $z/\Lambda \rightarrow \infty$, local turbulence conditions are completely decoupled from direct influence of the surface. As a consequence the height z should drop from the list of scaling parameters. It implies that all locally scaled variables (such as the Richardson number) should approach a constant value in this limit. This result will prove to be useful in the next section when we need a closure assumption for the equations of the stable boundary layer.

Dynamics

We have already seen that in a stable boundary layer the length scale of turbulence ℓ is much smaller than the boundary-layer height h. As we have noted during our treatment of the neutral boundary layer, this means that the necessary condition for the applicability of K-theory is satisfied. In other words, the dynamics of the stable boundary layer is again dominated by diffusion. Similar as for the neutral boundary layer, we find for the mean time scale

$$\mathcal{T}_m = \frac{h^2}{K}$$

Let us assume that this mean time scale is proportional to the forcing time scale ($\mathcal{T}_f$) for which we take the rotation time scale, f^{-1}. Next we must estimate the K. According to local scaling, discussed in the previous section, the most logical estimate would be $K \simeq (\tau/\rho)^{1/2}\Lambda$. However, both variables $(\tau/\rho)^{1/2}$ and Λ are still a unknown functions of height whereas we would prefer have a constant value for K. Therefore we take the values in the surface layer u_* and L so that $K \simeq u_* L$. We know that in stable conditions both u_* and L are small. Consequently the K in a stable boundary layer is small or mixing is slow.

Substituting the estimate found for K and using $\mathcal{T}_m \simeq f^{-1}$ we find for the height of the stable boundary layer

$$h = c\sqrt{\frac{u_* L}{f}}. \tag{4.29}$$

where c is a proportionality constant the value of which will be discussed below. This expression is known as the Zilitinkevich height.

One may wonder under which conditions expression (4.29) for the boundary-layer height is valid. Given our assumption that the time scale $\mathcal{T}_m \simeq \mathcal{T}_f$ we must conclude that the mean structure of the stable boundary layer is non-stationary. By integrating the equations that describe this mean structure as a function of time over the height of the boundary layer, we can find an evolution equation for the boundary-layer height (Nieuwstadt and Tennekes, 1981) which reads

$$\frac{dh}{dt} = -\frac{1}{T_d}(h - h_{eq}). \tag{4.30}$$

The stable boundary-layer height is thus described by an relaxation equation. The h as function of time approaches an equilibrium height h_{eq}. It turns out this this equilibrium height is equal to the Zilitinkevich expression (4.29). However, we can also learn another fact about the stable boundary layer from equation (4.30). Namely, the time scale T_d is given by

$$T_d \sim \left(-\frac{1}{\Delta\Theta}\frac{dT_0}{dt}\right)^{-1}$$

where dT_0/dt is the mean temperature change at the surface and $\Delta\Theta$ is the mean temperature difference across the boundary layer. Typical values for both dT_0/dt and $\Delta\Theta$ are: 1 °C/hour and 5 °C, respectively. This means that $T_d \simeq 5$ hours. In other words the development of the stable boundary layer is slow which agrees which the small mixing mentioned above.

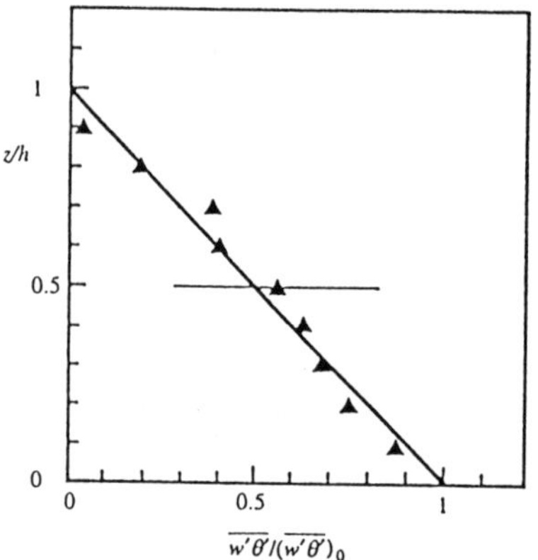

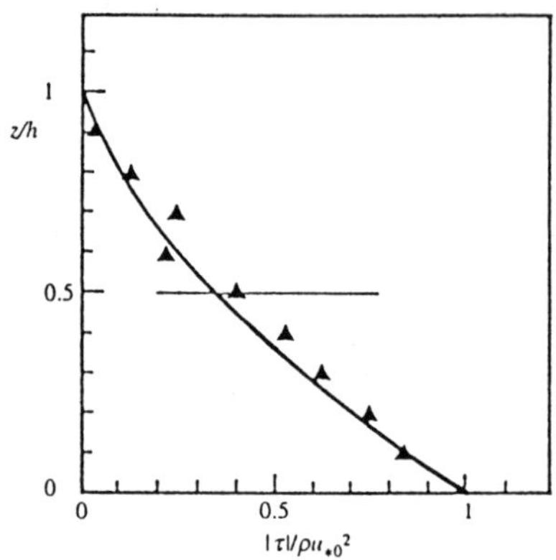

Figure 4.13: Vertical profiles of the temperature flux $\overline{w\theta}$ (left) and the momentum flux τ (right) in the quasi-stationary stable boundary layer; the symbols denote an average over a number of experiments; the variation between the experimental results is given by the horizontal line; the solid lines are the closed-form expression discussed in the text.

To end this section on dynamics of the stable boundary layer we consider its structure in somewhat more detail. Let us restrict ourselves to a so-called quasi-stationary boundary layer, which is defined here as a boundary layer in which turbulence is independent of time. This will be a quite exceptional case because we have seen above that in general the stable boundary layer is non-stationary. Furthermore, we shall assume that turbulence in our stable boundary layer can be described in terms of z-less scaling. For the Richardson number (4.28) this implies

$$Ri = Ri_{cr} \simeq 0.2$$

where the value follows from experiments. This result can be applied as a closure assumption in the equations that describe the stable boundary layer. It turns out that under the assumptions of quasi-stationarity these equations can be solved analytically (Nieuwstadt, 1984a) and (Garratt, 1992; pp 165-170). The results show that the Zilitinkevich height (4.29) is the height of this quasi-stationary boundary layer which agrees with its definition as equilibrium height mentioned above. Moreover, we find that the constant c in (4.29) is given by

$$c = \left(\sqrt{3}\, k\, Ri_{cr}\right)^{1/2} \simeq 0.37.$$

The values which follow from expression (4.29) with this value of c are in good agreement with observed boundary-layer heights (Nieuwstadt, 1984b).

With some representative values for u_* and L, say 0.2 m/s and 25 m respectively, we obtain from (4.29) that $h \approx 80$ m. We may thus conclude that compared to the convective boundary layer, discussed in section 4.4.3, the stable boundary layer is indeed quite shallow.

The solution of the equations for the quasi-stationary stable boundary layer also leads to explicit expression for the profiles of the temperature flux, $\overline{w'\theta}$ and the momentum flux, τ, which we have introduce above as scaling parameters. The result reads

$$\begin{aligned} \overline{w\theta} &= \overline{w\theta}_o \left(1 - \frac{z}{h}\right) \\ \frac{\tau}{\rho} &= u_*^2 \left(1 - \frac{z}{h}\right)^{3/2} \end{aligned} \qquad (4.31)$$

These expressions are compared with experimental observations in 4.13. The agreement is quite reasonable.

Dispersion

Based on our discussion of stable boundary-layer dynamics it will be clear that vertical dispersion in this case can be parameterized in terms of K-theory.

We first consider the dispersion from a surface source ($z_s = 0$). Applying local and z-less scaling and also using the expressions (4.4) and (4.5) for the velocity gradient and the diffusion coefficient, we find

$$\begin{aligned} K_h &\simeq k\, Ri_{cr}\, u_* L \sim 0.08 u_* L \\ \frac{dU}{dz} &\simeq 13 \frac{u_*}{L} \end{aligned} \qquad (4.32)$$

as representative values for the exchange coefficient and the velocity gradient in the lower part of the stable boundary layer.

The case of an instantaneous surface source is described by (4.11). For the dispersion coefficient σ_z given by expression (4.15) we then find with (4.32)

$$\sigma_z = \sqrt{2K_z t} \simeq 0.4\sqrt{u_* L t}. \qquad (4.33)$$

For a continuous surface source described by (4.14) we may derive for σ_z using (4.32)

$$\sigma_z \simeq \left(\frac{K_z x}{dU/dz}\right)^{1/3} \simeq 0.18 \left(L^2 x\right)^{1/3}. \qquad (4.34)$$

We thus find that the vertical dispersion of an instantaneous source and a continuous source are quite different. The explanation lies in the influence of strong velocity gradient.

For an elevated source we can use K-theory except in the initial phase when the σ_z is smaller than the turbulent length scale. For the profile of the K-coefficient we propose

$$K_z \simeq k\, Ri_{cr}\, u_* L \left(1 - \frac{z}{h}\right)^2$$

which follows from our solution of the quasi-stationary stable boundary layer discussed in the previous section. Nieuwstadt (1984b) and Robson (1987) have studied diffusion in the stable boundary layer based of this expression for the K-profile.

For a further discussion on dispersion in the stable boundary layer we refer to Hunt (1982) and Venkatram (1988).

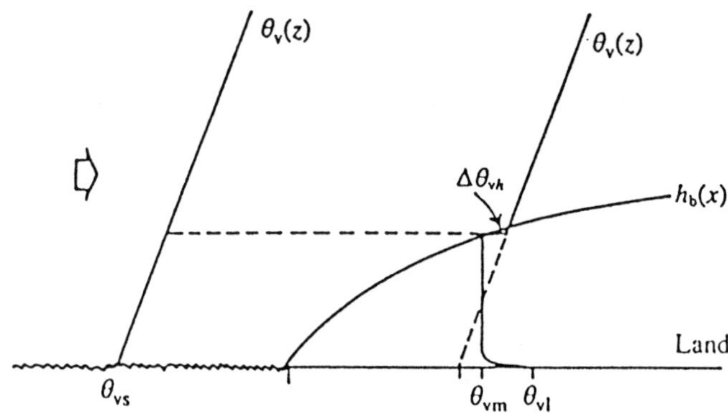

Figure 4.14: Schematic representation of a boundary layer that is advected from a surface over which $\overline{w'\theta'}_o \approx 0$ to a surface over which $\overline{w'\theta'}_o > 0$ (Garratt, 1992)

4.5 Heterogeneous boundary layer

In section 4.4 we restricted ourselves to the horizontally homogeneous boundary layer, i.e. a boundary layer without any horizontal variation. Here, we will consider the modification of the boundary layer when we allow inhomogeneous conditions. Let us assume that these conditions can be described in terms of a length scale: L_x. We may now distinguish two cases

The first case is: $L_x < h$, i.e. the length scale L_x is smaller than the boundary-layer height. Typical examples are heterogeneous terrain conditions, small scale topography or small scale boundary layer clouds. As result of the condition $L_x < h$ we may expect that the turbulence in these boundary layer can be directly influenced by the inhomogeneity. As a result, our assumption of quasi-stationary turbulence introduced in 4.4 may no longer be valid. As an example of this case we treat here the effect on the boundary layer due to a discontinuous change of surface boundary conditions, in particular the surface temperature flux and the stress.

In the other case, $L_x > h$, the boundary-layer height is smaller than the scale on which horizontal inhomogeneities occur. A typical example is large scale topography, such as hills. Here, we may expect that the turbulence structure is not directly influence by the inhomogeneity. This implies that locally the turbulence structure of a homogeneous boundary layer is preserved. Only the dynamics of the mean velocity and temperature will be influenced which is called mesoscale dynamics. It means that circulation patterns in the boundary layer on the scale L_x are generated and these influence the velocity and temperature profile. As an example we shall discuss the mesoscale dynamics of the convective boundary layer under the influence of large-scale variations of the surface boundary conditions.

4.5.1 Discontinuous surface boundary conditions

Temperature flux

First we consider the case of a boundary-layer which is advected from a surface over which $\overline{w'\theta'}_o \approx 0$, e.g. a sea surface, to a surface over which $\overline{w'\theta'}_o > 0$, e.g. a land surface. This case is illustrated in Fig. 4.14. It occurs for instance under sea breeze conditions. So at the coast

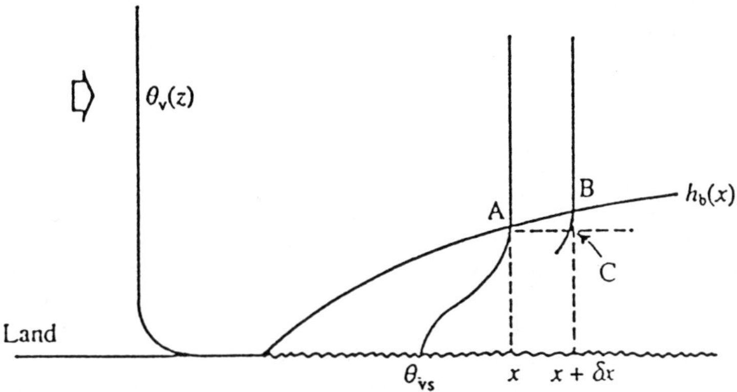

Figure 4.15: Schematic representation of a boundary layer which is advected from a surface over which $\overline{w'\theta'}_o \geq 0$ to a surface over which $\overline{w'\theta'}_o < 0$ (Garratt, 1992)

$(x = 0)$ we will have a discontinuous change of temperature flux boundary conditions.

We assume that the temperature profile over the sea can be characterized by a stable gradient, γ. When the land surface is reached a convective boundary layer will start to grow in this stable layer as a function of distance x from the coast. The time scale of this change can be characterized as

$$\mathcal{T}_{advection} = x/U$$

where U is the mean velocity perpendicular to the coast line. Let us consider the boundary-layer development at some distance from the coast where the convective boundary layer has grown to a height, h. Using the results on convective boundary layer dynamics discussed in section 4.4.3 we may estimate

$$\mathcal{T}_{turb} \sim \mathcal{T}_{mean} < \mathcal{T}_{advection}.$$

This implies that the mean structure of the stable boundary layer can be considered as quasi-stationary so that the development of h as function of x can be described by equations similar to (4.19) and (4.20) where t is replaced be x/U. The solution of these equations for the case of a constant $\overline{w'\theta'}_o$ is

$$h^2 = 2\frac{1+2\beta}{\gamma}\overline{w\theta}_o\frac{x}{U}.$$

The opposite case occurs when a convective boundary layer is advected from a land surface over a colder sea surface. The $\overline{w'\theta'}_o$ then changes discontinuously to a negative value and a stable boundary layer develops as a function of x from the coast. This case is illustrated in figure 4.15. If we omit again values of x close to zero and if we take into account the dynamics of the stable boundary layer discussed in section 4.4.4, we may estimate

$$\mathcal{T}_{turb} < \mathcal{T}_{mean} \sim \mathcal{T}_{advection}$$

where $\mathcal{T}_{advection}$ is again x/U with U the mean transport velocity. This means that our gradient diffusion approach for the stable boundary-layer development can be applied so that

$$h^2 \sim Kt \sim u_* L\,\frac{x}{U} \qquad (4.35)$$

The variable u_* and L, which are not always available can be rewritten in terms of other parameters. For instance, Garratt (1992) gives an alternative expression which reads

$$h^2 \sim 2.0\,10^{-4} U^2 \left(\frac{g\Delta\Theta}{T}\right)^{-1} x$$

where $\Delta\Theta$ is the temperature jump across the boundary layer. If we put in this expression some realistic values, say $U \approx 5$ m/s and $\Delta\Theta \approx 5$ °C, we find that the stable boundary layer grows very slow as a function of distance x. This again agrees with the weak mixing in a stable boundary layer which we have already encountered in section 4.4.4.

We have seen in section 4.4.3 and 4.4.4 that the convective boundary layer is in general much deeper than the stable boundary layer. So, when the convective boundary layer is advected over the sea surface some of the convective turbulence will end up above the stable layer. This turbulence is cut off from its buoyancy source and consequently it will decay.

Let us consider this decay process. First, it will be clear that turbulence is this part of the boundary layer can no longer be considered as quasi-stationary, i.e. production and destruction processes are no longer in equilibrium. As a matter of fact the production is completely stopped. So the development of turbulence will be governed by the time scale $\mathcal{T}_t$ and details of turbulence itself.

Nieuwstadt and Brost (1986) has studied this decay process by means of a large-eddy simulation. They consider a convective boundary layer in which at $t = 0$ the surface heat flux is set equal to zero. In Fig 4.16 we show the behaviour of two components of velocity fluctuations and of the temperature fluctuations as a function of dimensionless time

$$\frac{t}{\mathcal{T}_t} = \frac{t w_*}{h}.$$

The w_* is defined according to (4.16) in which the surface heat flux of the initial conditions is used. In our case we can interpret this dimensionless time as a dimensionless distance $w_* x/(hU)$. We find that the temperature fluctuations decay quite quickly and that within one time scale $\mathcal{T}_t \simeq h/w_*$ most of the temperature fluctuations have disappeared. The velocity fluctuations persist for almost one time scale and only then start to decay with the vertical velocity fluctuations disappearing faster than the horizontal velocity fluctuations. However, after $2-3$ time scales most fluctuations have dissipated. With representative values for w_*, h and U of 1 m/s, 1000m and 5 m/s respectively, we may estimate that it takes about 15 km for the convective turbulence to decay.

Surface stress

A spatial variation in the surface stress is frequently connected to a change in the surface roughness z_0. Here we shall only consider a discontinuity in the surface roughness. A schematic picture of such discontinuity in z_0 is shown in Fig 4.17.

We can distinguish two cases: smooth-to-rough and rough-to-smooth. In figure 4.18 we show for both cases the ratio, where τ_{o2} is the surface stress after the discontinuity which varies as a function of distance x and τ_{o1} the surface stress before the discontinuity. As may be expected, τ_{o2}/τ_{o1} is larger than 1 for the smooth-to-rough case and smaller than 1 for the rough-to-smooth case. However, near the point of discontinuity, i.e. $x = 0$, we see a larger overshoot in the ratio τ_{o2}/τ_{o1}. Is is caused by the fact that due to the sudden discontinuity, turbulence is completely out of equilibrium. As a consequence this behaviour can not be simulated be quasi-stationary turbulence models such as K-theory and more complicated turbulence models are required.

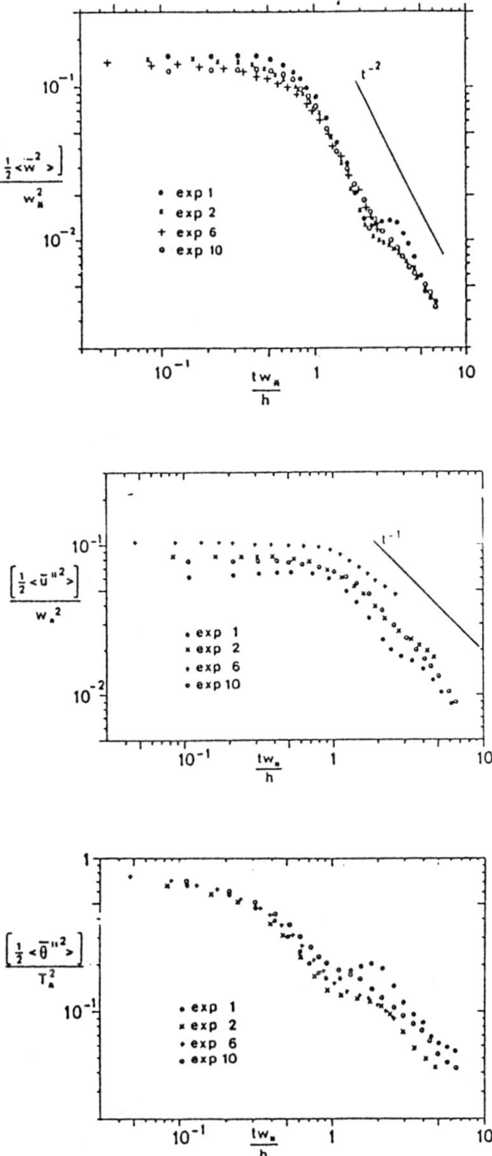

Figure 4.16: The decay as function of non-dimensional time of the vertical (top) and horizontal velocity fluctuations (center) and of the temperature fluctuations (bottom) for the case of a convective boundary layer in which at $t = 0$ the surface heat flux is set equal to zero (the data are averaged over the boundary layer); the symbols denote results of several large-eddy simulation runs (Nieuwstadt and Brost, 1986).

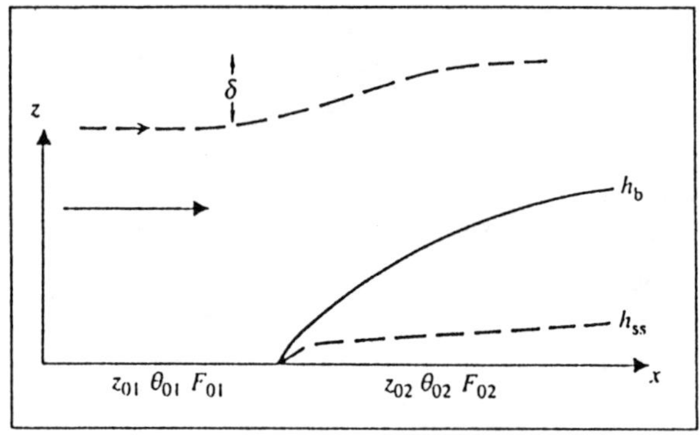

Figure 4.17: Schematic representation of a boundary layer under the influence of a discontinuous change of surface stress from z_{01} to z_{02} (Garratt, 1992)

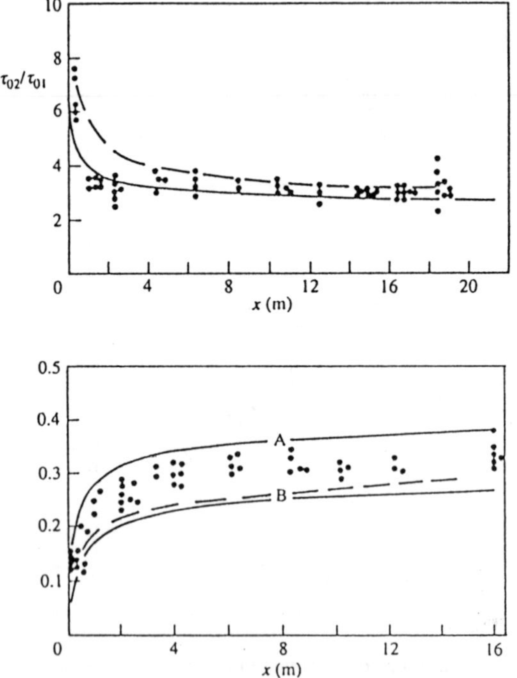

Figure 4.18: The surface stress after the discontinuity (τ_{o2} scaled with the surface stress before the roughness discontinuity (τ_{o1} as a function of distance from the discontinuity; top is the smooth-to-rough case and bottom the rough-to-smooth case; the lines are the results calculated with turbulence models and dots are observations (Garratt, 1992)

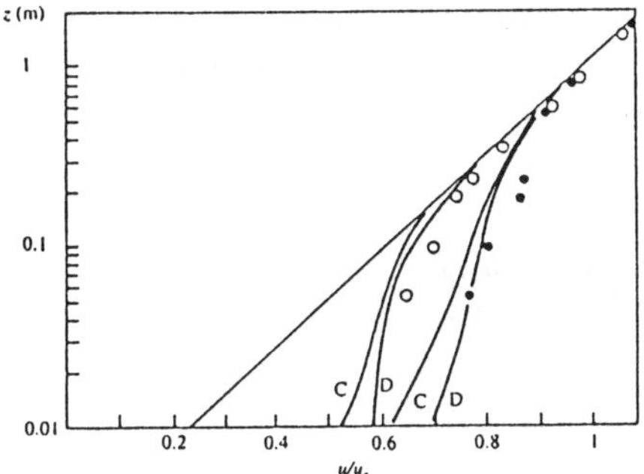

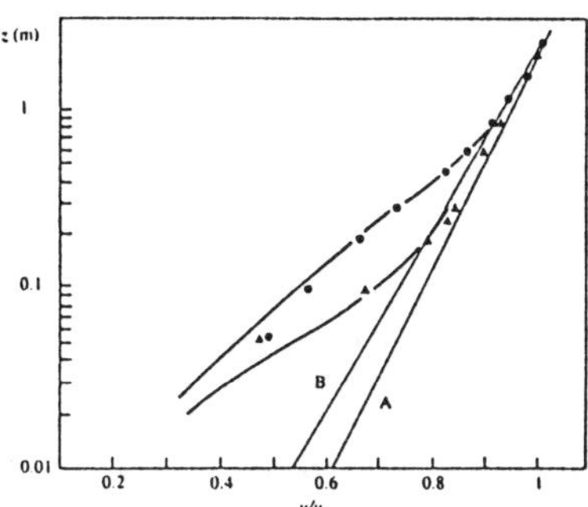

Figure 4.19: The change in velocity profile as a function of distance after the surface roughness discontinuity; left is the smooth-to-rough case and right the rough-to-smooth case; the lines labelled by letters are the results calculated for various values of the values of the roughness length z_0 and dots are observations; (Garratt, 1992)

Next we turn to the velocity profile. In Fig. 4.19 we show how this profile reacts to the roughness change. Before the discontinuity the profile is completely in equilibrium and can be described by the logarithmic profile (4.10). After the discontinuity the profile should eventually adjust to a new logarithmic profile which is representative for the new roughness length. The height to which the change in the velocity profile has progressed, is defined as the internal boundary-layer depth h_b. The development of h_b as a function of distance can be approximated as a solution of the equations which describe the growth of the boundary-layer over a flat plate (Garratt, 1992; p 110). This leads to the following expression

$$\frac{h_b}{x} = 0.9 \left(\frac{x}{z_o}\right)^{-0.2} . \tag{4.36}$$

In another approach to the development of h_b as a function of x we assume that the influence of the roughness can be modelled as a diffusion process. For the case of a neutral surface layer we have found that $\overline{z}$ (4.13) describes the height to which the influence of the diffusion process extends. With $\overline{z} \simeq 0.3 h_b$ we then find

$$\frac{dh_b}{dt} = U(h_b)\frac{dh_b}{dx} \simeq 3\, k u_* \tag{4.37}$$

where we have assume that $U(h_b)$ is the representative advection velocity.

If we take for $U(z)$ the logarithmic profile given by (4.10) the solution of (4.37) reads

$$\left(\frac{h_b}{z_o}\right)\left(\ln(\frac{h_b}{z_o}) + 1\right) \simeq \frac{x}{z_o} \tag{4.38}$$

which is an implicit equation for h_b as a function of x.

As mentioned above the h_b is the height of the internal boundary layer in which the influence of the roughness change can be felt. Within this internal boundary layer the velocity profile develops and is thus not in equilibrium. This means that the profile can not be described by a logarithmic function. However, very close to the surface the velocity can be considered in equilibrium and thus given be a logarithmic profile. The height of this so-called equilibrium layer is designated by h_{ss} and it can be approximated by

$$h_{ss} = 0.1 h_b. \tag{4.39}$$

With the equations (4.36) and (4.37) it follows that $h_b/x \approx 0.1$ so that $h_{ss}/x \approx 0.01$. So to obtain a logarithmic layer of 10 m height we have to be at least at distance of 1 km from the surface discontinuity.

4.5.2 Meso-scale Dynamics

We now turn to the case when the length scale L_x of the horizontal inhomogeneities is larger than the boundary-layer height. We consider here the influence of such meso-scale perturbations on the development of a convective boundary layer. For horizontally homogeneous conditions we have already discussed this boundary-layer type in section 4.4.3.

Let us repeat the equations which govern the horizontally homogeneous mixed-layer in a slightly different notation:

$$\begin{aligned} \frac{\partial D}{\partial t} &= W_e = \frac{cF_s}{\Delta} \\ \frac{\partial \Theta}{\partial t} &= \frac{(1+c)F_s}{D} \\ \frac{\partial \Delta}{\partial t} &= \Gamma W_e - \frac{\partial \Theta}{\partial t} \end{aligned} \tag{4.40}$$

where the surface temperature flux is here denoted be F_s and the depth of the mixed layer by D. We use upper case letters to indicated that this is the base state on which we shall superpose perturbations. The solution of these base state equations has already been given in (4.22).

Let us now apply perturbations to forcing of this boundary layer. First we consider the surface temperature flux, which we write as

$$f_s = F_s + f_s'.$$

The f_s is thus a spatially varying temperature flux which is written as a perturbation f_s' on the homogeneous flux F_s.

The second perturbation to the forcing is due to a height variation of the surface, i.e. topography, which is indicated as

$$s'$$

These perturbations in the forcing will influence all boundary-layer variables. We shall write this influence as a perturbation around the base state defined in (4.40) so that

$$\begin{aligned} &d' = d - D,\ \theta' = \theta - \Theta,\ u' = u, \\ &h' = d' + s',\ \delta' = \delta - \Delta,\ \gamma' = \gamma - \Gamma \end{aligned}$$

In Fig. 4.20 we have illustrated this configuration.

Next we assume that these perturbations are small with respect to the base state. Introducing them in the equations which govern the horizontally inhomogeneous boundary layer

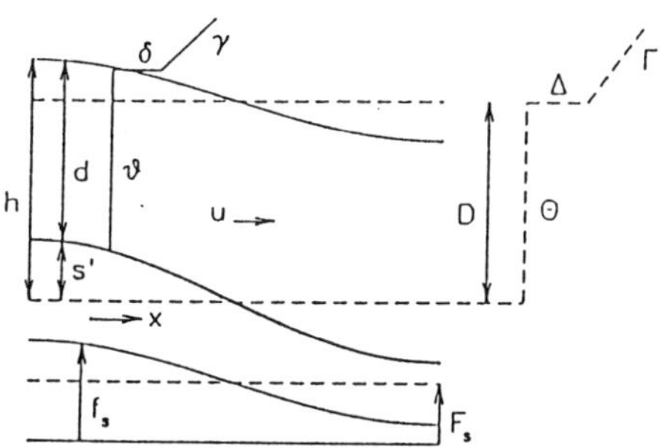

Figure 4.20: Schematic representation of the convective boundary layer under the influence of spatially varying temperature flux f_s and of variations in topography s. (Nieuwstadt and Glendening, 1989)

and linearizing the resulting equations we find

$$\begin{aligned}
\frac{\partial d'}{\partial t} &= -D\frac{\partial u'}{\partial x} + w'_e \\
\frac{\partial u'}{\partial t} &= F'_x - \frac{g}{T_0}\Delta\frac{\partial d' + s'}{\partial x} + \frac{g}{2T_0}D\frac{\partial \theta'}{\partial x} \\
\frac{\partial \theta'}{\partial t} &= \frac{(1+c)f'_s}{D} - \frac{(1+c)F_s}{D^2}d' \\
\frac{\partial \delta'}{\partial t} &= \gamma' W_e + \Gamma w'_e - \frac{\partial \theta'}{\partial t} \\
w'_e &= c\frac{f'_s}{\Delta} - c\frac{F_s}{\Delta^2}\delta'
\end{aligned} \tag{4.41}$$

where F'_x is a force due to motions induced by perturbations in the stable layer overlying the mixed-layer. It is parameterized as

$$F'_x = \frac{g}{T_0}L_p\frac{\partial \theta'}{\partial x}$$

where L_p is called the extinction length. It should be mentioned that in the derivation of (4.41) we assumed that the hydrostatic condition is valid. For further details regarding these equations and their derivation we refer to Nieuwstadt and Glendening (1989).

To find a solution of the linear set of equations (4.41) we assume that all perturbation can be written as horizontal wave solutions. This means for instance

$$d' = \hat{d}e^{ikx}, \cdots$$

where $\hat{d}$ is the amplitude of the wave and $k = 2\pi/L_x$ the wave number. Let us take for the horizontal wave length $L_x = 50$ km so that $k = 1.26\,10^{-4}$m. This clearly satisfies our condition that the scale of the perturbation is larger than the boundary-layer height. For the time dependence of the amplitude we write

$$\frac{\hat{d}}{D} \propto e^{\lambda(\tau - \tau_o)}$$

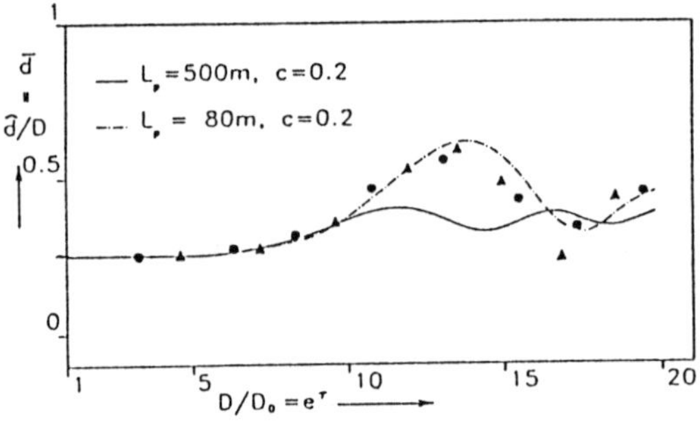

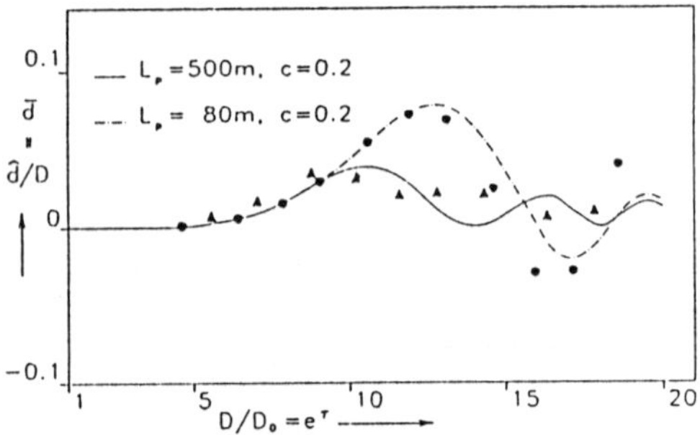

Figure 4.21: The dependence of the amplitude of the boundary-layer height perturbations as a function of D/D_0 for forcing by an inhomogeneous surface heat flux (top) and by topographical height perturbations (bottom); the line indicate solutions of the linearized set of equations discussed in the text; the symbols are computation with a full non-linear meso-scale model (Nieuwstadt and Glendening, 1989)

with

$$\tau = \ln\left(\frac{D}{D_0}\right).$$

We thus assume that the perturbations scaled with the base state solution are a power law function of D/D_0 where D_0 is the initial boundary layer depth. It is clear that for $\lambda > 0$ perturbations grow as a function of time and that they decay when $\lambda < 0$. It turns out that in the initial phase of the boundary layer growth, say $D/D_0 < 5$, values of λ can indeed become positive (Nieuwstadt and Glendening, 1989).

The time development of the perturbation is illustrated in Fig. 4.21. There we present a solution of the set of equations (4.41) for two values of the extinction length L_p. An example is given for both forcing conditions, i.e inhomogeneous temperature flux and topography. We see that initially the perturbation grow and can become quite large. For the case of the inhomogeneous surface heat flux the perturbation can reach a magnitude of half the base-state boundary-layer height D. After having reached a maximum the time behaviour of the perturbations becomes somewhat complicated. We have also compared in Fig. 4.21 the solution of the linearized equations (4.41) with the results of a full non-linear meso-scale model. The agreement between both models is quite acceptable which indicated the the meso-scale boundary-layer dynamics can be well described by linearized equations. However, the main result is that we have found that meso-scale perturbation can have a considerable influence on the development and structure of the convective boundary layer.

References

[1] Arya, S.P. 1988 *Introduction to micrometeorology.* Vol 42. International Geophysics Series, Academic Press, New York, U.S.A.

[2] Andren A., Brown A., Graf J., Mason P.J., Moeng C.-H., Nieuwstadt F.T.M. and Schumann U. 1994. Large eddy simulation of a neutrally stratified boundary layer: a comparison of four computer codes. Submitted to *Q. J. R. Meteorol. Soc.*.

[3] Briggs G.A. 1988. Analysis of diffusion field experiments. In *Lectures on air pollution modelling* (eds. A. Venkatram and J.C. Wyngaard), American Meteorological Society, Boston, U.S.A., pp 62–117.

[4] Fleagle R.G. and Businger J.A. 1980. *An introduction to Atmospheric Physics.* Academic Press, Orlando, Fla., U.S.A.

[5] Holtslag A.A.M and Nieuwstadt, F.T.M. 1986. Scaling the atmospheric boundary layer.*Bound.-Layer Meteorol.* **36**, 201–209.

[6] Hinze O. 1975. *Turbulence.* Mc Graw-Hill, New York, U.S.A.

[7] Hunt J.C.R. 1982. Diffusion in the stable boundary layer. In *Atmospheric turbulence and air pollution* (eds. F.T.M. Nieuwstadt and H. van Dop), Atmospheric Science Library, Kluwer Academic Publishers, Dordrecht, the Netherlands, pp 231–272.

[8] Garratt, J.R. 1992. *The atmospheric boundary layer.* Cambridge atmospheric and space science series, Cambridge University Press, Cambridge, England.

[9] Grant A.L.M. 1992 The structure of turbulence in the near-neutral atmospheric boundary layer. *J. Atmos. Sci.* **49**, 226–239.

[10] Lamb R.G., 1982. Diffusion in the convective boundary layer. In *Atmospheric turbulence and air pollution* (eds. F.T.M. Nieuwstadt and H. van Dop), Atmospheric Science Library, Kluwer Academic Publishers, Dordrecht, the Netherlands., pp159–230.

[11] Mason P. 1989. Large eddy simulation of the convective atmospheric boundary layer. *J. Atmos. Sci.* **46**, 1492–1516.

[12] Mason P.J. and Thomson D.J. 1987. Large-eddy simulation of the neutral-static-stability planetary boundary layer. *Q. J. R. Meteorol. Soc.* **113**, 413–443.

[13] Mason P.J. and Derbyshire S.H. 1990. Large-eddy simulation of the stably-stratified atmospheric boundary layer. *Bound.-Layer Meteorol.* **53**, 117–162.

[14] Nieuwstadt F.T.M. and Tennekes H. 1981. A rate equation for the nocturnal boundary-layer height. *J. Atmos Sci.* **38**, 1418-1428.

[15] Nieuwstadt F.T.M. 1984a. The turbulent structure of the stable boundary layer. *J. Atmos. Sci.* **41**, 2202-2216.

[16] Nieuwstadt F.T.M. 1984b. Some aspects of the turbulent stable boundary layer. *Bound. Layer Meteorol.* **30**, 31-54.

[17] Nieuwstadt, F.T.M. and van Dop H. 1982. *Atmospheric turbulence and air pollution.* Atmospheric Science Library, Kluwer Academic Publishers, Dordrecht, the Netherlands.

[18] Nieuwstadt F.T.M. and Brost R.A. 1986. The decay of convective turbulence.*J. Atmos. Sci.* **43**, 532–546.

[19] Nieuwstadt F.T.M. and Glendening J.W. 1989. Mesoscale dynamics of the depth of a horizontally non-homogeneous well-mixed boundary layer. *Beitr. Phys. Atmosph.* **62**, 275–288.

[20] Nieuwstadt F.T.M, Mason P.J., Moeng C.-H. and Schumann U. 1992. Large-eddy simulation of the convective boundary layer: a comparison of four computer codes. In *Turbulent Shear Flows 8*, Springer-Verlag, Berlin, pp 343–367.

[21] Reynolds W.C 1990. The potential and limitations of direct and large-eddy simulations. In *Whither turbulence? Turbulence at the crossroads.* (ed. J.L. Lumley) Springer-Verlag, pp 313–343.

[22] Robson R.E. 1987. Turbulent dispersion in a stable layer with quadratic exchange coefficient. *Bound. Layer Meteorol.* **39** 207–218.

[23] Schmidt H. and Schumann U. 1988. Coherent structure of the convective boundary layer derived from large eddy simulation. *J. Fluid Mech.* **200**, 511–562.

[24] Schumann U. 1989. Large-eddy simulation of turbulent diffusion with chemical reactions in the convective boundary layer. *Atmos. Env* **23**, 1713–1727.

[25] Schumann, U. 1991. Simulations and parameterizations of large eddies in convective atmospheric boundary layers. In *ECMWF-Workshop on Fine-scale modelling and the development of parameterization schemes, Reading 16-18 September.*

[26] Schumann U. and Friedrich R. 1987. On direct and large-eddy simulation of turbulence. In *Advances in Turbulence.* (eds. G. Comte-Bellot and J. Mathieu), Springer-Verlag.

[27] Schumann U. and Moeng C.-H. 1991. Plume fluxes in clear and cloudy convective boundary layers. *J. Atmos. Sci.* **48**, 1746–175.

[28] Sorbjan Z. 1989. *Structure of the atmospheric boundary layer.* Prentice Hall, Englewood, New Jersey, U.S.A.

[29] Stull, R. 1988. *An introduction to boundary-layer meteorology.* Atmospheric Science Library, Kluwer Academic Publishers, Dordrecht, the Netherlands.

[30] Tennekes H. and Lumley J.L. 1972. *A first course in turbulence.* The MIT Press, Cambridge, Massachusetts, U.S.A.

[31] Venkatram A. 1988. Dispersion in the stable boundary layer. In *Lectures on air pollution modelling* (eds. A. Venkatram and J.C. Wyngaard), American Meteorological Society, Boston, U.S.A., pp 229–258.

[32] Venkatram A. and Wyngaard J.C. 1988. *Lectures on air pollution modelling*American Meteorological Society, Boston, U.S.A.

[33] Weil J.C. 1988. Dispersion in the convective boundary layer. In *Lectures on air pollution modelling* (eds. A. Venkatram and J.C. Wyngaard), American Meteorological Society, Boston, U.S.A., pp 167–222.

[34] Willis G.E and Deardorff J.W 1976. A laboratory model of diffusion into the convective planetary boundary layer. *Q. J. R. Meteorol. Soc.* **102**, 427–445.

[35] Willis G.E and Deardorff J.W 1981. A Laboratory study of dispersion from a source in the middle of the convectively mixed layer. *Atmos. Env.* **15**, 109–117.

[36] Wyngaard, J.C. 1988. Structure of the PBL. In *Lectures on air pollution modelling* (eds. A. Venkatram and J.C. Wyngaard), American Meteorological Society, Boston, U.S.A., pp 9–61.

[37] Wyngaard, J.C. 1992. Atmospheric turbulence. *Annu. Rev. Fluid Mech.* **24**, 205–233.

[38] Wyngaard J.C. and Brost R.A 1984. Top-down and bottom-up diffusion of a scalar in the convective boundary layer. *J. Atmos. Sci.* **41**, 102-112.

[39] Wyngaard, J.C., Coté O.R. and Y. Izumi 1971. Local free convection, similarity and the budgets of shear stress and heat flux. *J. Atmos. Sci.* **28**, 1171–1182.

Address of the author

F. T. M. Nieuwstadt
Delft University of Technology
J. M. Burgers Centre, Lab. Aero and Hydrodynamics
Rotterdamseweg 145
2628 AL Delft, The Nederlands

V Some topics in turbulent diffusion

Han van Dop

5.1 Lagrangian particle diffusion models

5.1.1 Statistical approaches in turbulent dispersion

5.1.1.1 Elementary statistics. Consider a stochastic variable: X(t), which for example denotes the position of a randomly moving fluid particle, with an associated probability density function P(x), with the following properties

$$\begin{aligned} &P(x) \geq 0, \\ &\int P(x)dx = 1 \\ &\overline{X^n} = \int P(x)x^n dx. \end{aligned} \tag{5.1}$$

A stochastic process can in general then be defined as the function: Y(X,t). Note that the function may depend on more than one stochastic variable.

Next we define the probability density (function) as

$$P_k(y_1,t_1;y_2,t_2;y_3,t_3;y_4,t_4;\ldots\ldots y_k,t_k) \tag{5.2}$$

which is the probability that the particle takes the subsequent position y_1, y_2, ... at times t_1, t_2, ... respectively. Similarly we denote

$$P_{n-k,k}(y_1,t_1,\ldots y_k,t_k | y_{k+1},t_{k+1},\ldots y_n,t_n) \tag{5.3}$$

as the conditional probability density being the probability that a particle takes the positions y_{k+1}, y_{k+2}, ...with *assumed* previous positions y_1, y_2, ... at the corresponding times. It is now straightforward to infer from the above defintions the property:

$$P_3 = P_2(y_1,t_1;y_2,t_2)\, P_{1,2}(y_1,t_1;y_2,t_2 | y_3,t_3) \tag{5.4a}$$

and any other similar combination. Now a Markov Processes can be defined as

$$P_{1,n-1}(y_1,t_1\ldots y_{n-1},t_{n-1} | y_n,t_n) = P_{1,1}(y_{n-1},t_{n-1} | y_n,t_n), \tag{5.4b}$$

i.e. the conditional probability depends only on the previous location of the particle. From this property and the above definitions follows immediately the Chapman-Kolmogorov Equation:

A. Gyr and F-S. Rys (eds.), Diffusion and Transport of Pollutants in Atmospheric Mesoscale Flow Fields, 129–143.

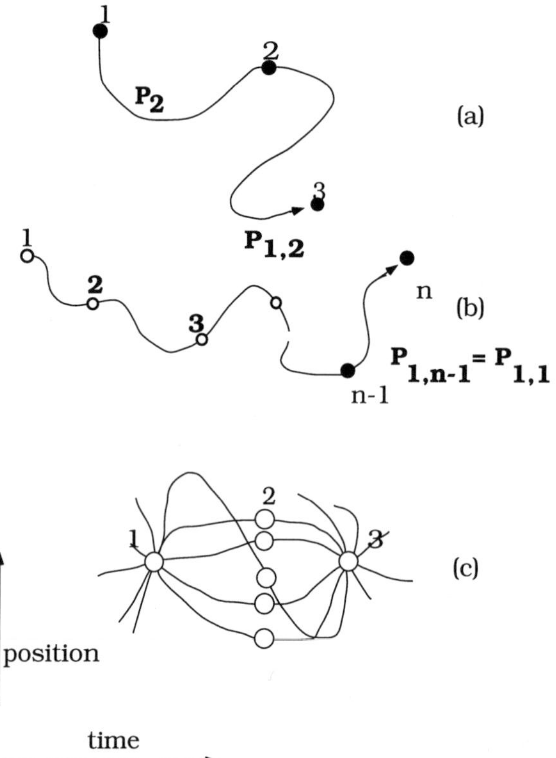

Figure 5.1. Illustration of some definitions in the text using fluid-particle trajectories. (a) the conditional probability density $P_{1,2}$ that a particle passes through position *3*, having passed through positions *1* and *2*; (b) the "Markovian property": the probability that a particle passes through position *n* having passed through position *n-1* is independent of previous particle positions *1,2,3, ...*; (c) a 'visualisation' of the Chapman-Kolmogorov equation.

$$P_{1,1}(y_1,t_1|y_3,t_3) = \int dy_2\ P_{1,1}(y_2,t_2|y_3,t_3) P_{1,1}(y_1,t_1|y_2,t_2) \tag{5.5}$$

5.1.1.2 The Master Equation If we assume stationarity:
$P(y_2) \equiv P_{1,1} = P_{1,1}(y_1|y_2;\tau); \quad (\tau = t_1 - t_2)$
and define the transition probability per unit time as: $W(y_2,y_1) = P_{1,1}/\tau$, we get from Eq. (5.5):

$$\frac{\partial P(y,t)}{\partial t} = \int \{W(y'|y)P(y',t) - W(y|y')P(y,t)\}\, dy' \tag{5.6a}$$

or in the discrete case:

$$\frac{dp_n(t)}{dt} = \sum_{n'} \{W_{nn'}p_{n'} - W_{n'n}p_n\} \tag{5.6b}$$

which are called master equations (see Van Kampen, 1981).

5.1.1.3 The Fokker-Planck equation. If we make a further assumption that the transition probability goes sufficiently fast to zero for increasing step size (W~0 for $\Delta y > \varepsilon$) and that W and P are slowly varying functions, we obtain after a Taylor series expansion assuming homogeneity ($W(y|y') = W(y', \Delta y), \quad (\Delta y = y - y')$):

$$\frac{\partial P}{\partial t} = -\frac{\partial}{\partial y}\{a_1 P\} + \tfrac{1}{2}\frac{\partial^2}{\partial y^2}\{a_2 P\},$$

$$with \quad a_n = \int (\Delta y)^n \, W(y, \Delta y) \, d\Delta y \tag{5.7}$$

the Fokker-Planck equation.

5.1.2 Applications In Turbulent Diffusion

5.1.2.1 The symmetric random walk. The symmetric random walk follows from the master equation if only neighbouring steps are permitted:

$$W_{nn'} = \frac{1}{\Delta t}\left\{\tfrac{1}{2}\delta_{n',n+1} + \tfrac{1}{2}\delta_{n',n-1}\right\}. \tag{5.8a}$$

Upon substitution in the (discrete) master equation (5.6b), we get:

$$\frac{\partial p}{\partial t} = K\frac{\partial^2 p}{\partial y^2} \quad \text{with} \quad K = \lim_{\substack{\Delta y \to 0 \\ \Delta t \to 0}} \tfrac{1}{2}\frac{\Delta y^2}{\Delta t} \tag{5.8b}$$

which can be identified as the diffusion equation.

5.1.2.2 The dichotomic Markov process. A better simulation of turbulent motion can be obtained by assuming that particles move with uniform, finite velocity $\pm u$ and have probability a of velocity change per unit of time. Denoting the probability densities that a particle moves in the positiveand negative direction by $p^{\pm}$respectively, we can formulate the master equation for this process as:

$$\frac{dp_n^+}{dt} = \overset{(1)}{\frac{1-\alpha}{\Delta t} p_{n-1}^+} + \overset{(2)}{\frac{\alpha}{\Delta t} p_{n+1}^-}$$

$$\frac{dp_n^-}{dt} = \overset{(3)}{\frac{1-\alpha}{\Delta t} p_{n+1}^-} + \overset{(4)}{\frac{\alpha}{\Delta t} p_{n-1}^+} \tag{5.9}$$

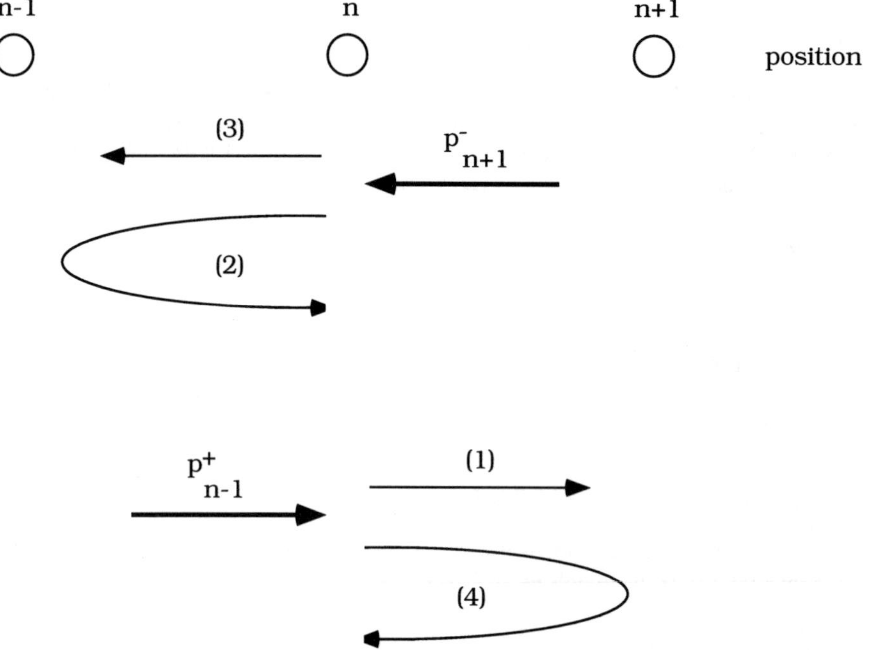

Figure 2. Illustration of the dichotomic one-dimensional Markov process. Particles move with uniform velocity u in either positve or negative y-directions. Positions $n-1$, n and $n+1$ with spacing $u\Delta t$ are depicted. After a timestep Δt (thin lines) particles may change direction. The numbers in parenthese refer to the corresponding terms in Eqs.(5.9).

(the terms at the right-hand side are illustrated in Fig. 2). Making Taylor expansions of $p^{\pm}_{n\pm1}$:

$$p^{\pm}_{n\pm1} = p^{\pm} \pm \Delta y \frac{\partial p^{\pm}_{n}}{\partial y} + \ldots\ldots,$$

and defining flux and particle concentration (C)as $u(p^{+} - p^{-})$ and $(p^{+} + p^{-}) / \Delta y$ respectively, we obtain

$$\frac{\partial^2 C}{\partial t^2} + \frac{2\alpha}{\Delta t}\frac{\partial C}{\partial t} + (2\alpha - 1)u^2 \frac{\partial^2 C}{\partial y^2} = 0.$$

When the velocity-change frequency $a(= \alpha / \Delta t)$is introduced we obtain in the limiting case

$$\Delta t, \Delta y, \alpha \to 0,$$
$$\alpha / \Delta t \to a$$
$$\Delta y / \Delta t = u$$

the Telegraph equation:

$$\frac{\partial^2 C}{\partial t^2} + 2a\frac{\partial C}{\partial t} - u^2 \frac{\partial^2 C}{\partial y^2} = 0 \tag{5.10}$$

(Monin and Yaglom, 1971). From (5.10) it follows that for a -> 0 pure advection results. If now the following limit is taken:

$$\left.\begin{array}{l} u \to \infty \\ a \to \infty \\ u^2 / a \to K \end{array}\right\},$$

the diffusion equation (5.8b) is retained. Note that (5.10) has the "correct" limits for the variance of the particle distribution for large and small times and that it reproduces Richardson's law. These can be derived from (5.10) by defining the moments of the particle distribution $\overline{y^n} = \int y^n C(y,t)dy$ and choosing suitable initial conditions.

5.1.2.3 Spectral diffusion Theory. If we take as a starting point that the transition probability depends on the step size only: $W_{nn'} = W(|n - n'|)$, this yields

$$\frac{\partial p}{\partial t} = \int W(|y - y'|)\left\{p(y') - p(y)\right\} dy' \tag{5.11}$$

which, taking its Fourier transform gives:

$$\frac{\partial \tilde{p}}{\partial t} = \left\{\tilde{W}(k) - 1\right\} \tilde{p}(k), \qquad \text{with } \tilde{p} \equiv \int p(y)\, e^{iky} dy \quad etc.$$

5.1.3 The Langevin equation

5.1.3.1 Application of stochastic methods in homogeneous turbulence. Photographs of the tracks of fluid elements in turbulent flows' show that they are mainly curved trajectories connected by sudden changes of direction. Such observations immediately suggest that an instructive way to model and study these random trajectories is by calculating the displacements of fluid elements or particles whose accelerations and initial velocities are random processes. These are called random flight models. Because there are some similarities between these trajectories and those of microscopic particles under the action of Brownian motion the changes (*dW*) in the velocity *W* (say in the z direction) of marked fluid elements are, in homogeneous conditions, most simply modeled by the Langevin equation,

$$dW = -(W / T_L)dt + a_2^{1/2} d\omega_t. \tag{5.12}$$

In Eq. (5.12), *W(t)* denotes the (vertical) component of the three dimensional particle velocity. The coefficient a_2 is to be specified by the physics of the process. The random variable $d\omega_t$ represents the (rapid) random acceleration, which is uncorrelated from one timestep to the next, while T_L is the time scale over which the fluid element's velocity is correlated while it undergoes the random acceleration. This is obviously analogous to the velocity of the "Brownian" particle under the action of molecular forces.

If we are interested in a timescale of the order of T_L, the nature of the acceleration during the sudden jumps in velocity does not matter, so $d\omega_t$ is modeled as a random increment such that

$$\begin{aligned} &\overline{d\omega_t} = 0, \\ &\overline{d\omega_t^2} = dt \\ &\overline{W d\omega_t} = 0 \\ &\overline{W = 0} \end{aligned} \qquad (5.13)$$

where the overbar denotes an ensemble mean. In homogeneous turbulence the solution to Eq. (5.12) is simple. Noting that T_L and a_2 are constant it can be solved with the initial condition $W(t = 0) = w_r$, where w_r is drawn from a Gaussian distribution with mean zero and variance equal to the fluid velocity variance $\sigma_w^2 \left(\equiv \overline{u_3^2}\right)$. After squaring and averaging the solution we get the well-known results

$$\overline{W^2(t)} = \left(\sigma_w^2 - \tfrac{1}{2} a_2 T_L\right) e^{-2t/T_L} + \tfrac{1}{2} a_2 T_L. \qquad (5.14)$$

Since $\overline{W^2(\infty)} = \sigma_w^2$ we have $a_2 = 2\left(\sigma_w^2 / T_L\right)$. Note that multiplying (5.12) with $W(t')$ and averaging yields an exponential correlation function:

$$R_L(t / T_L) = e^{-t/T_L}.$$

There are two main problems that arise in the use of such a model to predict the statistics of displacements of fluid elements in a given turbulent flow. First, there is no theory which specifies T_L and the statistics of dw_t, in terms of the turbulent velocities measured at a number of fixed (or moving) points. Second, the basic Langevin equation (5.12) is incorrect if the turbulence is inhomogeneous and unsteady. In such a flow an ensemble of marked fluid elements may have a mean acceleration, and also there may be a mean rate of increase of the mean square velocity of fluid elements, which is equivalent to a correlation between the fluctuating velocity and the acceleration.
Compared to calculations of the mean square of particle displacements using the diffusion equation with a turbulent diffusivity K(z), random flight models give a better description of dispersion for times less than about T_L. Also for $t > T_L$ they are able to include more properties of the turbulence than the usual K-theory formulation.
Random flight models can be solved numerically quite easily. Moreover, they are exactly mass conserving. Statistically stable solutions, however, are obtained only when the trajectories of thousands of particles are evaluated. This is now well feasible with present day pc capacities.

5.1.3.2 Application of stochastic methods in non-homogeneous turbulence. The application of (5.12) in homogeneous turbulence is well established. The more interesting applications are, however, in non-homogeneous (and non-stationary) turbulence
The first computations of dispersion by random flight techniques in non-homogeneous turbulence used the basic Langevin equation and they were restricted to problems where

only the time scale T_L and not the energy of the turbulence varied with position. In the cases where the fluid velocity variance $\overline{u_3^2}$ varies with position, a mean acceleration equal to $\partial \overline{u_3^2}/\partial z$ needs to be added to the random walk model. Its introduction has either been based on some simple arguments that it prevented unphysical accumulation of particles in regions of low variance, or on an asymptotic analysis of the Langevin equation. In the above approach it was intuitively assumed that the Langevin equation could be applied by letting T_L or s_W (and thus a_2) be a function of Z. The expression for a_2, however, was derived from an asymptotic analysis of the homogeneous case, where the coefficients in the Langevin equations were constants. This analysis does not apply in the general case where a_2 is a function of the particle's position and therefore it is not a priori clear that its functional form should be the same as in the homogeneous case.
In this field important progress was been made by Thomson (1987) who developed consistent analyses based on certain general assumptions.

5.1.3.3 Application in the convective boundary layer. Dispersion in convective conditions was first studied by Willis and Deardorff (1976,1978,1981)by means of a water-tank . The characteristics of the dispersion as revealed in these experiments, were quite unexpected. Material emitted at ground level first remains at the surface but then rises quickly to mid-level of the boundary layer, whereas particles released from elevated stacks first descend and then move to mid-level. These characteristics have since been confirmed through observations in a wind-tunnel and in field experiments.
Eulerian K theory is unable to describe these phenomena, it does not give the proper vertical dispersion for small times. The rising of the plume would require negative K values, and in reality material emitted from elevated sources reaches the ground sooner and closer to the source than these models predict. An alternative is to use Lagrangian models in which the motion of individual fluid particles is considered. In these models the dispersion for times smaller than the Lagrangian time scale is better described. Besides that, conservation of mass is ensured and no problems occur with numerical stability of the equations used.
For dispersion in non-homogeneous conditions we consider the Langevin equation of the following form

$$dW = -(W / T_L)dt + d\mu, \tag{5.15}$$

where $d\mu$ is now a random function with moments which are all order dt:

$$\overline{d\mu^n} = a_n dt \qquad (n = 1,2,3...) \tag{5.16}$$

(De Baas et al.,1986). Expressions for a_1 and a_2 etc. were obtained by Thomson (1984) and Van Dop (1985). The first two read:

$$a_1 = \frac{\partial \overline{u_3^2(z)}}{\partial z} dt$$

$$a_2 = \left\{ \frac{2\overline{u_3^2(z)}}{T_L} + \frac{\partial \overline{u_3^3(z)}}{\partial z} \right\} dt \tag{5.17}$$

Usually during daytime, air is heated at the surface and the boundary layer becomes unstable and dominated by buoyantly driven turbulence. The vertical turbulence structure becomes organized in a pattern of updraughts and downdraughts, where on the average

the updraughts move faster than the downdraughts. Because the vertical speed, averaged over a large horizontal area, should be zero, the downdraughts occupy a larger area than the updraughts at each level of the boundary layer and the vertical velocity distribution is skew. If the turbulence is inhomogeneous, the vertical velocity distribution is height-dependent. The Eulerian properties of the convection (outside the surface layer, where the stresses are constant) can be scaled with the convective velocity w_* and the height, z_i of the boundary layer.
It is assumed that the Lagrangian and Eulerian correlation functions, $R_L(t)$ and $R_E(t)$ are similar in shape but displaced by a scale factor : $R_L(\beta t) = R_E(t)$ (Hanna, 1981). From the definition of T_L it follows that $\beta = T_L / T_E$ The Lagrangian properties then also scale with w_* and z_i. We get

$$0.24 z_i / w_* < T_L < 0.55 z_i / w_* \qquad (5.18)$$

but we have to keep in mind that there is a large uncertainty in the constants.
Many measurements have been carried out to determine the profiles of the second and third moments of the vertical turbulence velocity.

5.1.3.4 Alternative formulations in non-homogeneous turbulence. There are, however, several problems in applying (5.15) in non-homogeneous conditions. The first is that it does not satisfy the so-called "well-mixed condition". What this means will be explained below.
In addition the above model formulation requires the first tree moments of the random increments to be O(dt) with higher moments $O(dt^2)$. Now any random variable X must satisfy[1]

$$\left(\overline{X^3}\right)^2 \le \overline{X^2}\,\overline{X^4} \qquad (5.19)$$

Hence there is no random forcing with the required moments. In some non-Gaussian cases the situation can be even worse with the model requiring increments with negative variance. Of course for a particular finite value of t it may be possible to choose the distribution of the random increments in order to satisfy the criteria at least approximately and this is the way (5.15) was successfully used. However it is rather unsatisfactory mathematically because the length of the time step Δt plays an essential role and the well-mixed condition is normally only satisfied approximately. Therefore other model formulations were proposed. A summary:
We can generally formulate the Langevin equation as

$$\begin{aligned} dW &= a\,dt + b\,d\mu \\ dZ &= W dt \end{aligned} \qquad (5.20)$$

where $d\mu$ are independent random velocity increments with a Gaussian distribution. The probability distribution function (PDF) of the tracer particles should satisfy the Fokker-Planck equation corresponding with (5.20), which in stationnary conditions is:

$$w\frac{\partial P}{\partial z} = -\frac{\partial}{\partial w}(aP) + \tfrac{1}{2}\frac{\partial^2}{\partial w^2}\left(b^2 P\right) \qquad (5.21)$$

---

[1] This can be easily proven by noting that $\int \left(\lambda x + x^2\right)^2 P(x)\,dx \ge 0$. The condition that the resulting quadratic expression in λ has no roots leads to the desired unequality (5.19).

The well-mixed condition requires that if the particles of tracer are initially well-mixed they will remain so. Therefore if $P_a(z,w)$) is the PDF of all fluid elements, it should also satisfy the Fokker- Planck Equation (5.21). Writing the latter as

$$aP_a = \frac{\partial}{\partial w}\tfrac{1}{2}\left(b^2 P_a\right) + \phi, \quad \text{with}$$
$$\frac{\partial \phi}{\partial w} + w\frac{\partial P_a}{\partial z} = 0 \tag{5.22}$$

we obtain the well-mixed condition in mathematical form.

In the homogeneous, stationary and one dimensional case with no mean flow a and b are according to (5.12) equal to $-W/T_L$ and $\left(2\sigma_w^2 / T_L\right)^{1/2}$ respectively.

Based on local isotropy it can be argued that b equals $\left(2C_0\varepsilon\right)^{1/2}$. An expression for a can then be derived from (5.22), where, for example, in non-homogeneous, convective turbulence various expressions can be used for P_a which reflect the particular properties of the CBL.

5.1.3.5 Boundary conditions. The inhomogeneous nature of convective turbulence is such that very small time steps (often just a few seconds) are necessary in random walk particle models to resolve the turbulence near the top and bottom of the mixed layer. Failure to observe the time step restriction leads to an accumulation or deficit of particles in the boundary regions. Various solutions has been proposed such as reflection conditions or adapting the turbulence profiles such that the boundaries become unattainable

5.1.3.6 Buoyant dispersion. It is clear that the concept of the random walk in order to describe turbulent transport of a tracer in a turbulent flow has some attractive features. Applications in non-buoyant dispersion, howevere, have been sparse. Zanetti and Al Madani (1984) and Cogan (1985) formulated simple Lagrangian models for buoyant dispersion. Here the possibility of describing turbulent motion of buoyant tracers in a more fundamental way will be summarized.
There are two major aspects that distinguish buoyant and passive dispersion:
(i) buoyant fluid particles "create" their own turbulent field in an environment which may be laminar or turbulent and (ii) the exchange processes between the plume particles and the (turbulent) environment should be included in the dynamics.
A review and a detailed analysis of entrainment and turbulent transfer between the plume and its environment was given by Netterville (1990).
We summarize the basic concepts. We shall define the plume as the volume which contains a mixture of ambient and (most of the) originally released, buoyant fluid. The envelope of the plume is the imaginary and in a way, arbitrary, boundary of this volume. Some of the original buoyant fluid may be taken away from the plume and become so remote that it is no longer considered to be part of it. On the other hand the volume of the plume expands due to turbulent intrusions of ambient air resulting in an increasing ambient fraction and consequently, a gradual loss of plume temperature and vertical acceleration (in a cooler and calm environment).
A Lagrangian plume particle can now be defined as a small entity which possesses the mean characteristics (velocity, temperature) of the plume. Stochastic fluctuations, directly related to the turbulent intensity within the plume, determine the rate of growth of the plume particles extrainment.

Ultimately the plume (particle) dynamics (described by temperature and velocity and not by heat content and momentum) must converge to the environmental dynamics.
This is only an initial step in the formulation of Lagrangian buoyant diffusion and will contain some assumptions and unsolved problems. However, as we shall see shortly, the model is able to explain the basic features of plume motion in a non-adiabatic environment quite well. Besides, it is indicated how environmental turbulent characteristics can be included. Also, the effect of varying thermal stratification of the boundary layer can be easily included.

Starting point is the basic Langevin equation (5.12) which is extended to include the temerature as another stochastic variable. We define the Lagrangian temperature of a plume particle by Q. Then according to inertial sub-range theory we expect the following relationship between the temperature dissipation function ε_B and the structure function

$$D_\Theta \equiv \overline{\left(\Theta(t) - \Theta(t+\tau)\right)^2} = \varepsilon_B \tau. \tag{5.23}$$

We expect that the autocorrelation of Q is exponential and, consequently, its spectrum behaves as ω^{-2}. Analogously to the dynamical equation (5.12), we put

$$d\Theta = -\frac{\Theta - \Theta_a}{T_B} dt + \varepsilon_B^{1/2} d\omega_B, \tag{5.24}$$

where ε_B is the temperature dissipation function and T_B is a timescale for the temperature relaxation. The random increment $d\omega_b$ has similar properties to $d\omega_t$(see Eqs. 5.13)
As in (5.12), where the first RHS term "models" the mean velocity decrease, the first RHS term in Equation (5.24) represents the decrease in temperature of the particle due to turbulent intrusion of (cooler) ambient air. We shall neglect other processes such as heat exchange by radiation or molecular motion.
It should be realized that this concept, where a plume is thought to consist of a superposition of many independent trajectories of plume particles, disregards all kinds of nonlinear processes within the infrastructure of the plume. Their dynamical effects are thus ignored here. The equations of motion can now be formulated including the buoyant acceleration in Equation (5.12) as

$$dW = -(W / T_L)dt + Bdt + a_2^{1/2} d\omega_t, \tag{5.25}$$

where B is defined by $B \equiv \frac{g}{T}(\Theta - \Theta_a)$. From Equation (5.24) and the definition of B we may formulate the dynamic behaviour of B as

$$dB = -\frac{B}{T_B} dt - N^2 W dt + \varepsilon_B^{1/2} d\omega_B, \tag{5.26}$$

where we have included the possibility that the environment is diabatic: $\frac{g}{T} d\Theta_a = N^2 W dt$, where N is the Brunt-Vaisala frequency. Together with the trajectory equation, dZ= Wdt, Equations (5.25-26) form the basic set of equations.

5.2 Chemical reactions in dispersion processes

5.2.1 Introduction

We consider the conservation laws for the instantaneous flow quantities, successively the continuity equation:

$$\frac{\partial \rho}{\partial t} + \frac{\partial}{\partial x_i} \rho u_i = 0 \text{ or} \tag{5.27}$$

$\frac{1}{\rho}\frac{d\rho}{dt} = -\nabla . \boldsymbol{u}$, or, approximately $\nabla . \boldsymbol{u} \cong 0$, and the momentum equation

$$\frac{\partial u_i}{\partial t} + u_j \frac{\partial u_i}{\partial x_j} = -\frac{1}{\rho}\frac{\partial P}{\partial x_i} - 2\varepsilon_{ijk}\Omega_j u_k - g\delta_{i3} + \nu \frac{\partial^2 u_i}{\partial x_j \partial x_j}. \tag{5.28}$$

For a chemical tracer the conservation equation is

$$\frac{\partial \rho c}{\partial t} + \frac{\partial \rho u_j c}{\partial x_j} = Q, \tag{5.29}$$

where Q denotes a source/sink term. The usual way to proceed is to decompose the Eulerian variables in the equations (5.27-29) in a mean and a fluctuating value and then take the ensemble average. In a horizontally uniform but vertically stratified atmosphere (5.29) then reduces to

$$\frac{\partial C}{\partial t} + U_j \frac{\partial C}{\partial x_j} = \frac{1}{\rho_0}\frac{\partial}{\partial x_j}\left(\rho_0 K \frac{\partial C}{\partial x_j}\right) + Q \tag{5.30}$$

where upper case symbols denote mean quantities, lower case fluctuating quantities and the overbar the average.

5.2.2 Closure

In an imperfect analogy with molecular diffusion, the turbulent flux term in (5.30) can be modelled by a gradient transfer term:

$$-\rho_0 \overline{u_j c} = K \frac{\partial C}{\partial x_j} \tag{5.31}$$

where the properties of the flow are incorporated in the eddycoefficient K,

and thus

$$\frac{\partial C}{\partial t}+U_j\frac{\partial C}{\partial x_j}=\frac{1}{\rho_0}\frac{\partial}{\partial x_j}\left(\rho_0 K\frac{\partial C}{\partial x_j}\right)+Q', \tag{5.32}$$

5.2.3 Formulations which include chemistry

5.2.3.1 The intensity of segregation. When species chemically react we have for each species a transport equation according to (5.32) and additional equations describing the chemical interactions. It should be noted that the latter are formulated in instantaneous concentrations which has important consequences as we shall see for the over all formulation. We consider simple systems such as

$$\begin{aligned} &NO_2+h\nu \overset{O_2}{\Leftrightarrow} NO+O_3\\ &O_3+h\nu \overset{H_2O}{\rightarrow} OH\\ &OH+NO_2\rightarrow HNO_3\\ &HNO_3+NH_3\Leftrightarrow NH_4NO_3 \end{aligned} \tag{5.33}$$

and which can be formulated in general as $A+B\rightarrow P$. The rate equation for the formation of p is $dp/dt=k\,a.b$. Decomposing this in mean and fluctuating quantities leads to

$$\begin{aligned} &\frac{dP}{dt}=k\,A.B(1+I_s), \quad with\\ &I_s=\frac{\overline{a'b'}}{A.B}, \quad \text{the intensity of segregation} \end{aligned} \tag{5.34}$$

This quantity can be obtained from a mathematical model (see e.g. Schumann, this volume) or from experimental observations (see Fig. 3).

5.2.3.2 Turbulence closure, K-theory and chemistry. However, also for the combination of turbulent transport and chemistry slightly different formulations are required. From (5.27-29) we infer the turbulent transport equation in the horizontally homogeneous case:

$$\frac{\partial\overline{wc}}{\partial t}=-\overline{w^2}\frac{\partial C}{\partial z}-\frac{1}{\rho_0}\overline{c\frac{\partial p}{\partial z}}-\frac{\partial\overline{w^2c}}{\partial z}. \tag{5.35}$$

Note that assuming stationarity, neglecting the third order correlation term and parameterizing the pressure term by $\overline{wc}/\tau$, a relationship is obtained between the flux and the mean concentration gradient, resulting in the parabolic diffusion equation (c.f. Eq. 5.32) with $K=\overline{w^2}\tau$.

Repeating the same analysis for the system

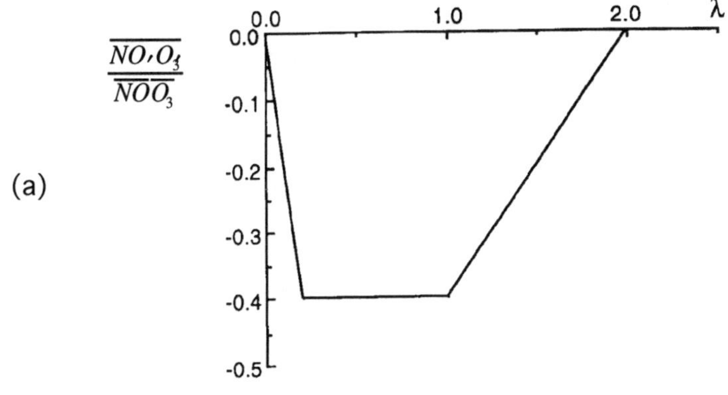

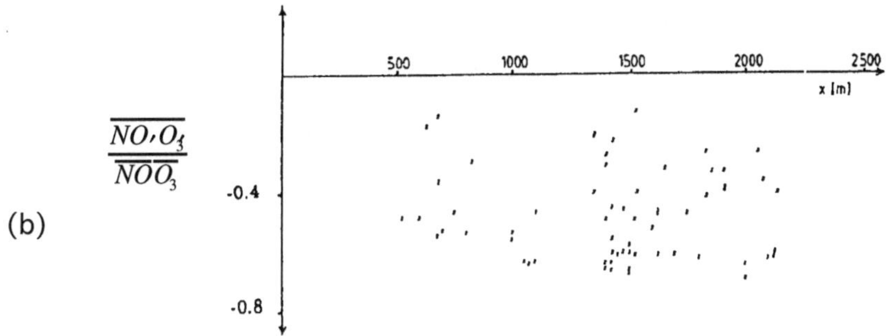

Figure 3. Intensity of segregation (a) modelled in a chemically reactive plume as a function of radial distance(λ) and (b) according to measurements collected in West Germany (by Frank et al.). Source: J. Vilà-Guerau de Arellano (1992).

$$\overset{(a)}{NO_2} + h\nu \underset{k_1}{\overset{k_2}{\rightleftarrows}} \overset{(b)}{NO} + \overset{(c)}{O_3}$$

$$\begin{aligned} \frac{\partial a}{\partial t} &= -k_2\, a + k_1\, b.c \\ \frac{\partial b}{\partial t} &= +k_2\, a - k_1\, b.c \\ \frac{\partial c}{\partial t} &= +k_2\, a - k_1\, b.c \end{aligned} \tag{5.36}$$

yields the following set of equations:

$$\overline{wa} = -\overline{w^2}\tau\frac{\partial A}{\partial z} + k_1\tau\left(\overline{wb}\,C + \overline{wc}\,B\right) - k_2\tau\overline{wa}$$

$$\overline{wb} = -\overline{w^2}\tau\frac{\partial B}{\partial z} - k_1\tau\left(\overline{wb}\,C + \overline{wc}\,B\right) + k_2\tau\overline{wa} \qquad (5.37)$$

$$\overline{wc} = -\overline{w^2}\tau\frac{\partial C}{\partial z} - k_1\tau\left(\overline{wb}\,C + \overline{wc}\,B\right) + k_2\tau\overline{wa}$$

which can be rewritten as

$$-\overline{wc_i} = R_{ij}\,\overline{w^2}\tau\frac{\partial C_j}{\partial z} \qquad (5.38)$$

with

$$R_{ij} = \frac{1}{D}\begin{pmatrix} 1 + k_1\tau B + k_1\tau C & k_1\tau C & k_1\tau B \\ k_2\tau & 1 + k_2\tau + k_1\tau B & -k_1\tau B \\ k_2\tau & -k_1\tau C & 1 + k_2\tau + k_1\tau C \end{pmatrix} \qquad (5.39)$$

with $D = 1 + k_2\tau + k_1\tau C + k_1\tau B$. The indices 1,2,3 in Eq. 5.38 correspond to A, B and C respectively (Vilà-Guerau de Arellano, J. et al., 1992a,b).
Within the framework of K-theory (5.38-39) are the correct expressions for the turbulent transport of chemically active tracers.

References

Cogan J.L. 1985 Monte Carlo simulations of buoyant dispersion. *Atmospheric Environment* **19**, 867-878.
Hanna, S.R. 1981 Lagrangian and Eulerian timescales in the daytime boundary layer. *J. Appl. Met.* **20**, 242-249.
De Baas A.F., van Dop, H. & Nieuwstadt, F.T.M. 1986 An application of the Langevin equation for inhomogeneous conditions to dispersion in a convective boundary layer. *Q. J. R. Met. Soc.* **112**, 165-180.
Monin A.S. & Yaglom A.M. 1975 Statistical Fluid Mechanics. M.I.T. Press.
Netterville, D.D.J. 1990 Plume rise entrainment and dispersion in turbulent winds. *Atmospheric Environment* **24A,** 1061-1081.
Thomson D.J. 1984 Random walk modelling of diffusion in inhomogeneous turbulence. *Q. Jl R. met. Soc.* **110**, 1107-1120.
Thomson D.J. 1987 Criteria for the selection of the stochastic models of particle trajectories in turbulent flows. *J. Fluid Mech.* **180**, 529-556.
Van Dop, H., Nieuwstadt, F.T.M. & Hunt, J.C.R. 1985 Random walk models for particle displacements in inhomogeneous unsteady turbulent flows. *Phys. Fluids* **28**, 16391653.
Van Kampen, N.G. 1981 Stochastic processes in physics and chemistry. North-Holland.
Vilà-Guerau de Arellano, J., Talmon, A. & Builtjes, P.J.H. 1990 A chemically reactive plume model for the NO-NO_2-O_3 system. *Atmospheric Environment* **24A**, 2237-2246.
Vilà-Guerau de Arellano, J. 1992a The influence of turbulence on chemical reactions in the atmospheric boundary layer. Ph. D. Thesis Universiteit Utrecht.
Vilà-Guerau de Arellano, J. & Duynkerke, P.G. 1992b Second-order study of the covariance between chemically reactive species in the surface layer. *J. Atmos. Chem.* In press.

Willis G.E. & Deardorff, J.W. 1976 A laboratory model of diffusion into the convective planetary boundary layer. *Q. Jl R. Met. Soc.* **102**. 427-445.
Willis G.E. & Deardorff, J.W. 1978 A laboratory study of dispersion from an elevated source within a modeled convective planetary boundary layer. *Atmospheric Environment* **12**, 1305-1311.
Willis G.E. &Deardorff, J.W. 1981 A laboratory study of dispersion from a source in the middle of the convective mixed layer. *Atmosheric Ennronment* **15**, 109-117.
Zannetti, P. & Al-Madani, N. 1984 Simulation of transformation, buoyancy and removal processes by Lagrangian particle methods. In *Proc. 14th Int. Technical Meetimg Air Pollution modeling and its Application* (ed. Ch. de Wispelaere), pp. 733-744, Plenum.

Dr. Han van Dop
Institute for Marine and Atmospheric Research (IMAU)
Utrecht University,
Princetonplein 5
P.O. Box 80005
3508 TA Utrecht
The Netherlands
e-mail: dop@fys.ruu.nl

VI. Modelling Diffusion and Dispersion of Pollutants

Torben Mikkelsen

6.1 Scales of Atmospheric Motion

Diffusion and dispersion processes in atmospheric turbulence naturally divide according to different time and space scales. It is common to distinguish scales of atmospheric motion according to:

The Planetary Surface Layer
The Planetary Boundary Layer
The Meso-γ scale (2-20 km)
The Meso-β scale (20-200 km)
The Meso-α scale (200-2000 km)

In spatial extension, the surface layer extends from the ground and up to a few hundred meters in the vertical. It is characterized by approximately constant fluxes of heat and momentum which can be determined by measurements near the surface.

Pielke (1984) assigns the mesoscale to features with a horizontal extension large enough that the hydrostatic approximation can still be valid, yet small enough that the larger (synoptic) scales geostrophic and gradient winds can be considered inappropriate as approximations to the actual wind circulation. Vertically, the surface layer extends a few hundred meters whereas the mesoscales extends throughout the entire troposphere.

Mesoscale models utilize grid spacing ranging between a few meters for modelling of local scale dispersion phenomena and out to 20 km or more in numerical weather forecast models utilized on the regional scale. The corresponding time scales ranges from fractions of a second in the near-surface layer concentration fluctuation modelling to tenth of hours for the meso-α scale weather system calculations.

Dispersion of pollutants in the mesoscale atmosphere behaves accordingly different on the various scales and is influenced by the initial source dimension, the heat and momentum content of the release, by nearby building effects, by downwind changes in roughness and heat fluxes (ie. the atmospheric stability) and by transition through time and spatially changing boundary layers. Additional terrain effects include heterogeneous surfaces, hills, rolling and steep complex terrain.

Mesoscale dispersion scenarios are in addition influenced by larger-scale differential heating (sea- and land breezes, up and down slope winds, valley circulations and orographically induced wind storms).

A. Gyr and F-S. Rys (eds.), Diffusion and Transport of Pollutants in Atmospheric Mesoscale Flow Fields, 145–164.

Modelling of mesoscale dispersion is commonly based on diffusion calculations relating to one or more of the following three different classes:

1) *Statistical theory*, including single and multiple particle releases and Monto Carlo-techniques
2) *Similarity theory*, and
3) *Gradient transport* (K-theories)

6.2 Review of Diffusion Theories

The paper intends to provide the reader with an overview of the most important methods and principles frequently applied to atmospheric dispersion modelling. The format is therefore chosen to be "encyclopedic" and chronological.

The volume of this chapter does not allow for a detailed derivation and formal presentation of each "index", therefore, each method is presented by a resume containing of the most important principles and formulas. The reader is encouraged to consult the suggested literature for a deeper and more thorough presentation and derivation of the various techniques and theories.

The paper follows the above classification starting with *Statistical Theory*.

6.2.1 Statistical Theory (G.I. Taylor's formula)

Taylor established, in his pioneering work from 1921, a relationship between the plume width (standard deviation σ) and the Lagrangian auto-correlation of the dispersing fluid, - which in this case is the turbulent atmosphere.

Define:

$\sigma_y^2(t)$ Mean square crosswind deviation from fixed axis at time t.
$v(t)$ Crosswind fluid or particle velocity in a homogeneous field of turbulence.
$R_L(\tau)$ The corresponding Lagrangian (single-particle) autocorrelation function $\langle v(t)v(t+\tau)\rangle$.

Absolute (fixed frame) single-particle diffusion: *Plumes.* G.I. Taylor related the growth rate of the average plume width σ, to the Lagrangian autocorrelation of the atmospheric turbulence. The formula applies to the ensemble-averaged plume width including plume meander, in a coordinate system fixed to the source point.

To distinguish Taylor diffusion from the turbulent diffusion of an instantaneously released multiple-particle puff, in which case the coordinate system moves with the puffs center of mass (see below), Taylor diffusion is often referred to as *absolute* or *fixed frame* or *single-particle* diffusion.

Taylor's formula for plume diffusion reads:

$$\frac{1}{2}\frac{d\sigma_y^2}{dt} = \int_0^t R_L(\tau)d\tau = \overline{v^2}\, t_a(t) \qquad (6.1)$$

where $<v^2>$ denotes the variance or the velocity (square) scale of the turbulence, and t_a denotes the corresponding *time scale* or *memory time* for the *diffusion process*.

Asymptotic limits. In the limit for large *travel times* or *diffusion times*, $t_a(\infty)$ defines the Lagrangian *integral time scale* t_L through:

$$\int_0^\infty R_L(\tau)d\tau = \overline{v^2}\, t_L \qquad (6.2)$$

Taylor's formula, equation (6.1), yields in this limit the corresponding *far field* solution:

$$\sigma_y^2 = 2\,\overline{v^2}\, t_L\, t \qquad (6.3)$$

In the opposite limit, with *travel times* or *diffusion times* much smaller than the Lagrangian integral time scale: $t \ll t_L$, the auto-correlation function becomes: $R_L(\tau) \sim 1$ and Taylors formula results in the so-called *near-field* solution:

$$\sigma_y^2 = \overline{v^2}\, t^2 \qquad (6.4)$$

Spectral form. By introduction of the Lagrangian *variance spectrum* $S_L(\omega)$:

$$\overline{v^2}\, S_L(\omega) = \frac{1}{2\pi}\int_{-\infty}^{\infty} R_L(\tau)\, e^{-i\omega\tau}\, d\tau \qquad (6.5)$$

Taylor's formula can be represented in its corresponding spectral form:

$$\frac{1}{2}\frac{d\sigma_y^2}{dt} = \overline{v^2}\, t^2 \int_{-\infty}^{\infty} S_L(\omega)\, \frac{\sin^2(\omega\tau/2)}{(\omega\tau/2)^2}\, d\tau \qquad (6.6)$$

Relative (moving frame) multiple-particle diffusion: *Puffs*. An analogous Taylor's formula for puff diffusion, relating the ensemble averaged puff standard deviation σ_p to some turbulence property of the flow, can be formulated in terms of *relative* fluid particle velocities $v_r(t)$.

Define:

$\sigma_p^2(t)$:	Mean square crosswind deviation about puffs center-of-mass.
$v_r(t)$:	Crosswind particle velocity relative to puffs center-of-mass velocity.
$R_r(t,\tau)$:	Lagrangian auto-correlation function for v_r: $<v_r(t)v_r(t+\tau)>$.

In these terms, Taylor's formula for relative diffusion then reads:

$$\frac{1}{2}\frac{d\sigma_p^2}{dt} = \int_0^t R_r(t,\tau)d\tau = \overline{v_r^2}\, t_r(t) \tag{6.7}$$

This equation applies to the ensemble-averaged instantaneous puff size relative to the puffs center-of-mass coordinate. To distinguish the spread in equation (6.7) from the diffusion of a plume (equation (6.1)), puff diffusion is often referred to as *relative*, *moving frame* or *two-particle* diffusion.

Moving frame diffusion (σ_p^2), plus center-of-mass meandering (X_{cm}^2), equals fixed frame diffusion (σ_y^2).

The relative velocity and time scales defined by Equation (6.7) can for Gaussian shaped puffs be modelled as (Mikkelsen et al., 1986):

$$\overline{v_r^2} = \overline{v^2} \int_{-\infty}^{\infty} S_E(k)\,(1 - e^{-k^2\sigma^2})\, dk \tag{6.8}$$

$$t_r(t) = \frac{1}{\overline{v_r^2}} \int_0^t R_L(\tau)d\tau \tag{6.9}$$

where $S_E(k)$ is the fixed point Eulerian covariance spectrum and $R_L(\tau)$ the single-particle Lagrangian auto-correlation function quantities.

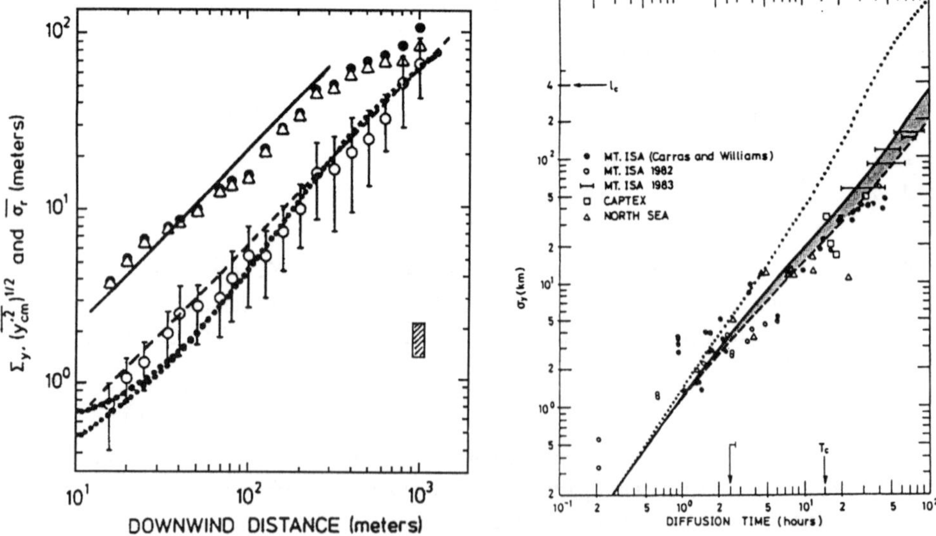

Figure 6.1 Relative and absolute diffusion on different scales. Surface layer ground level release (left) and Tropospheric cloud releases (right).

The left figure shows diffusion of ground level released smoke to distances out to 1 km from the source. The lower set of data points shows the observed puff diffusion whereas the upper set of points and represents plume diffusion including meander. The solid lines show model results based on equations (6.7)-(6.9) (lower) and on equation (6.1) (upper), and represents puff and plume diffusion respectively. The right Figure shows tropospheric puff diffusion data observed on the meso-α scale. Plume diffusion has here no meaning on this scale and consequently is only cloud or puff diffusion shown. The model results (lines) are calculated by the puff model in equation (6.7)-(6.9) (Mikkelsen & Eckman, 1983), (Mikkelsen et al., 1988).

6.2.2 Statistical Theory (Monte-Carlo Techniques)

Homogeneous turbulence

$v(t)$:	Crosswind turbulent particle velocity
$1/\beta$:	Lagrangian memory time scale t_L
$\rho_L(\Delta t)$:	Lagrangian auto-correlation e^{-T}, $T = \Delta t/t_L$
$a(t)$:	Random uncorrelated accelerations $<a(t_1)a(t_2)>$

Random force model - Langevin equation:

$$\frac{dv}{dt} + \beta v = a(t) \tag{6.10}$$

Corresponding Marcow equation:

$$v(t+\Delta t) = \rho_L(\Delta t)v(t) + \eta(t) \tag{6.11}$$

where $\eta(t)$ is a Gaussian process ($\mu=0$; $\sigma_\eta^2=(1-\rho^2)<v^2>$).

Solutions. Fixed frame - absolute diffusion (Taylor's formula):

$$\sigma_y^2 = 2\overline{v^2}t_L^2[T - (1-e^{-T})] \tag{6.12}$$

Moving frame - relative diffusion ($v_r(0)=0$) (Gifford, 1982):

$$\sigma_p^2 = 2\overline{v^2}t_L^2[T - (1-e^{-T}) - (1-e^{-T})^2] \tag{6.13}$$

i.e.: Moving frame diffusion = Fixed frame diffusion - meander

Inhomogeneous Gaussian turbulence

$P(w)$: Vertical (w) velocity pdf (Gaussian)

σ_w, t_L: functions of vertical coordinate z

Corresponding inhomogeneous Marcov equation (Wilson et al. (1983); Legg and Raupach (1982)):

$$w(t+\Delta t) = aw(t) + b\sigma_w\eta(t) + (1-a)t_L\, d\sigma_w^2/dz \tag{6.14}$$

where $a=\exp(-\Delta/t_L)$ and $b=(1-a^2)^{\cdot}$.

Inhomogeneous non-Gaussian turbulence (well mixed). (Thomson (1987), Luhar and Britter (1989), Tassone (1993))

P(z,w): Joint (non-gaussian) pdf of (z,w).

Fokker-Planck equation:

$$w\frac{\partial}{\partial z}P + \frac{\partial}{\partial w}(aP) = \frac{1}{2}\frac{\partial^2}{\partial w^2}(b^2 P) \tag{6.15}$$

The corresponding Marcov equations are obtained from:

1) Specify: $P^+(w)$ and $P^-(w)$ (skewed up/down-draft pdf).
2) Assume local isotropy (Kolmogorov subrange: $b=(c_o\varepsilon)^{\cdot}$).
3) Solve Fokker-Planck for a, then integrate the corresponding non-linear stochastic Marcow DE's:

$$dw = a(z,w)dt + b(z,w)d\eta \tag{6.16}$$

$$dz = wdt \tag{6.17}$$

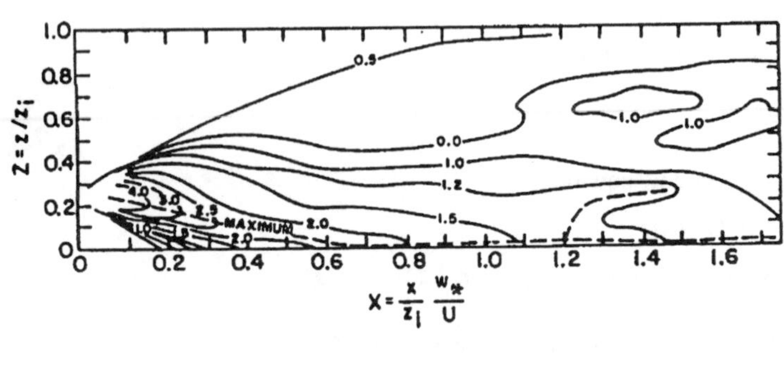

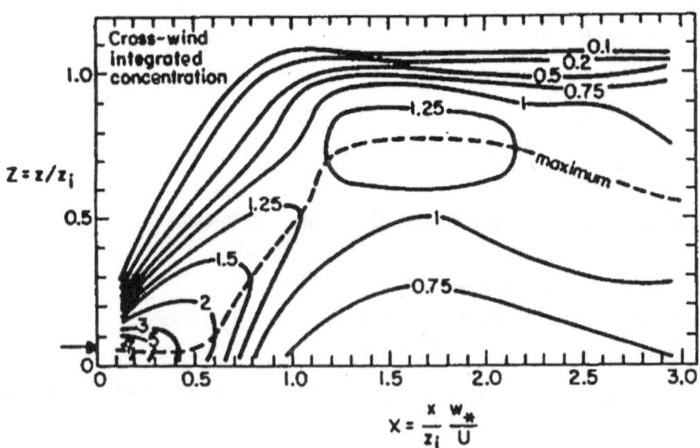

Figure 6.2. Deardorf/Willis water tank (Hanna, 1982)

6.2.3 Two-Particle Statistics, Distance Neighbour-function and Concentration Fluctuations

Concentration Mean $\langle c\rangle$ and Variance $\langle c^2\rangle$

Define

$\langle c\rangle$:	Mean concentration
$\langle c(y_1,t)c(y_2,t)\rangle$:	Concentration co-variance
$S(y')$:	Source distribution at time t_o
$P_1(y,t;y',t_o)$:	One-particle transition probability in time interval $t-t_o$.
$P_2(y_1,y_2,t;y'_1,y'_2,t_o)$:	Joint two-particle transition probability in time interval $t-t_o$.

Then, according to Batchelor (1952), Durbin (1984):

$$\overline{c}(y,t) = \int P_1(y,t;y',t_0)S(y')dy' \tag{6.18}$$

$$\overline{c^2}(y,t) = \iint P_2(y,y,t;y_1',y_2',t_0)S(y_1')S(y_2')dy_1'dy_2' \tag{6.19}$$

Note: $\langle c^2\rangle$ and thereby concentration fluctuations (c'), ($\langle c'^2\rangle = \langle c^2\rangle - \langle c\rangle^2$), relates to two-particle statistics!

6.2.4 Lidar Measurements of Plume Statistics

Crosswind concentration profiles (Jørgensen & Mikkelsen, 1993)

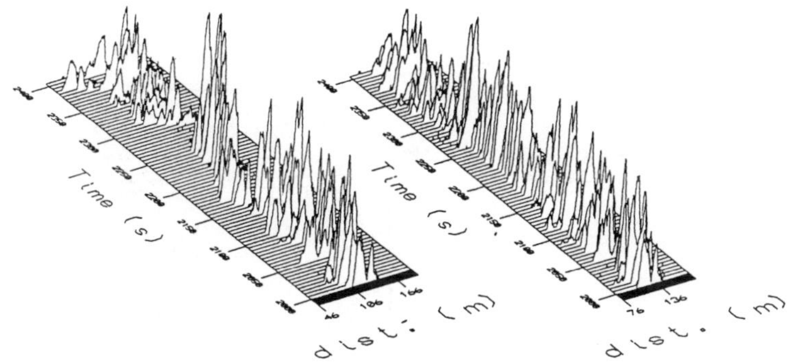

Figure 6.3. Lidar measurements: Fixed frame (left) and moving frame (right)

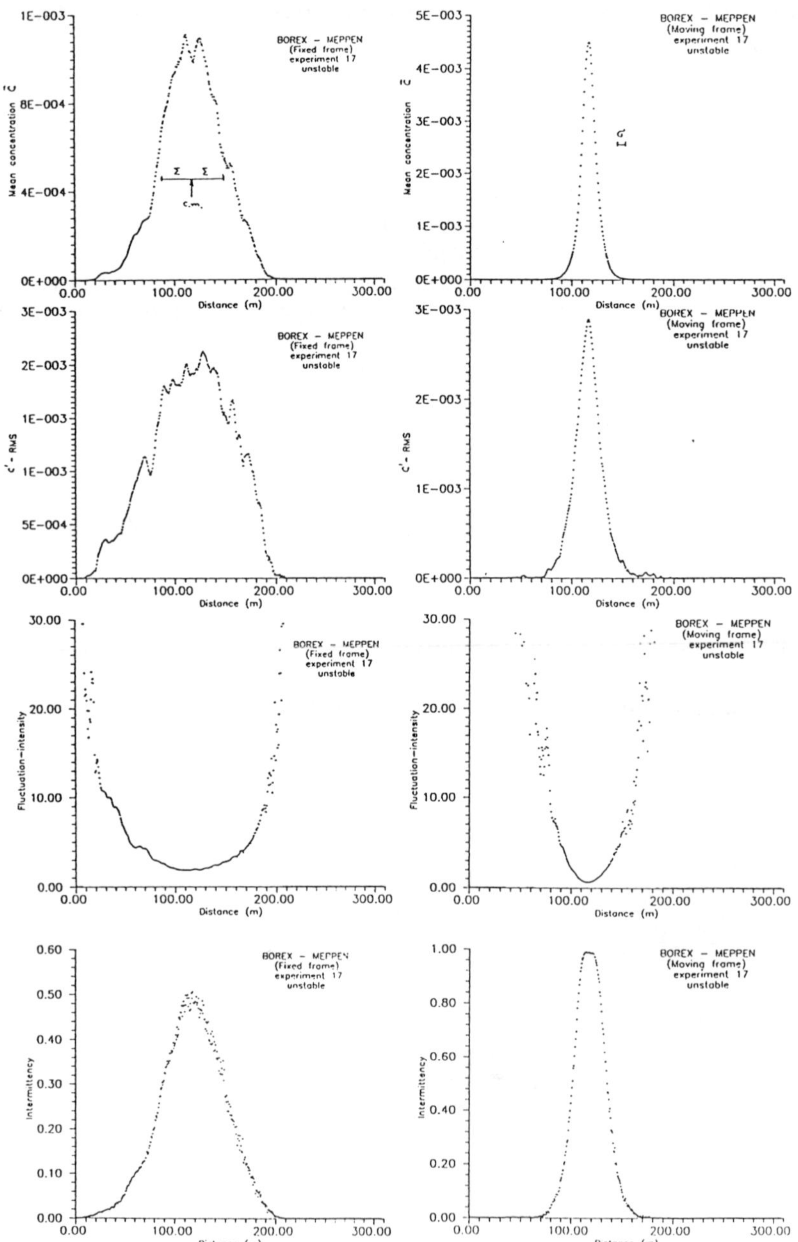

Figure 6.4. Crosswind plume statistics of <c>, $<c'>_{rms}$, Intensity and Intermittency: Left side: Fixed frame; Right side: Moving Frame

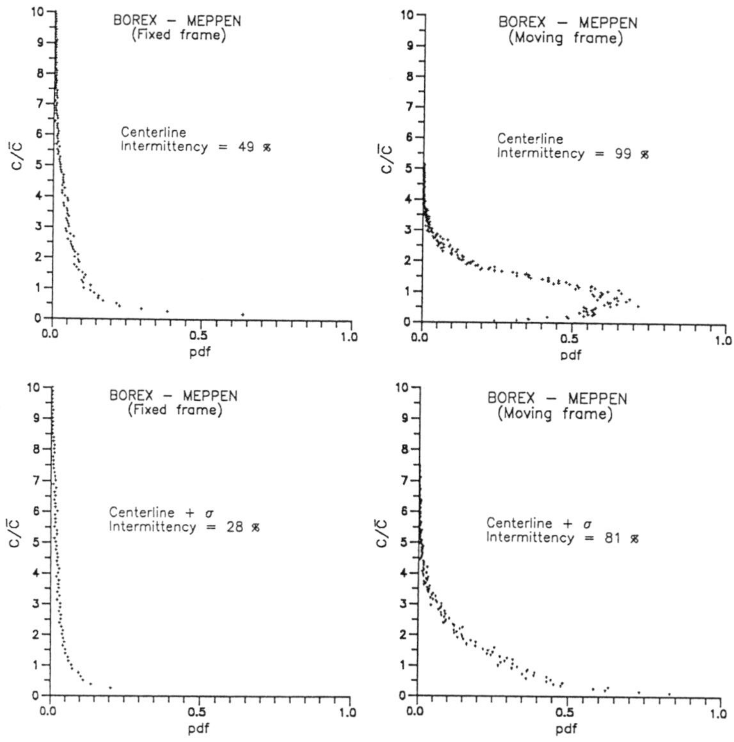

Figure 6.5. Measured concentration pdf's. Left side: Fixed Frame. Right side: Moving frame. Top: Centerline. Bottom: Plume margin (+1σ).

6.2.5 Distance-Neighbour Functions

Define: Distance Neighbour function (Richardson, 1926):

$$q(l) = \int_{-\infty}^{\infty} c(y)c(y+l)\, dl \tag{6.20}$$

Diffusion equation . for q(l):

$$\frac{\partial}{\partial t}q(l,t) = \frac{\partial}{\partial l}K(l)\frac{\partial}{\partial l}q(l,t) \tag{6.21}$$

Richardson, 1926: $K(l) \propto l^{4/3}$; (note: $\sigma_p^2 \propto \varepsilon T^3$),
Batchelor, 1952: $K(l) \propto \langle l^2 \rangle^{2/3}$; (note: $\langle l^2 \rangle = 2\sigma_p^2$).

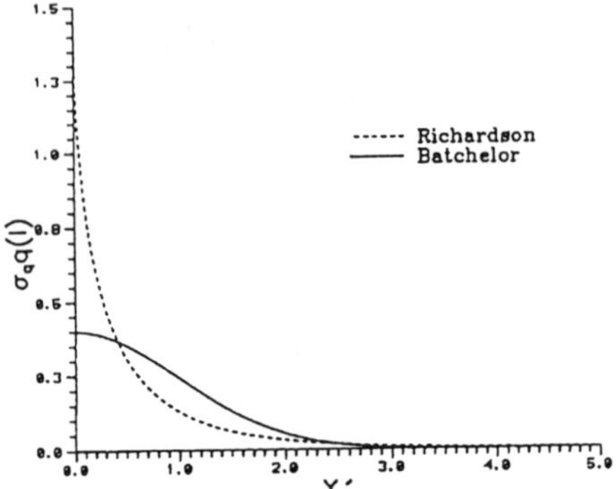

Figure 6.6. Solutions to q(l) diffusion eqation.

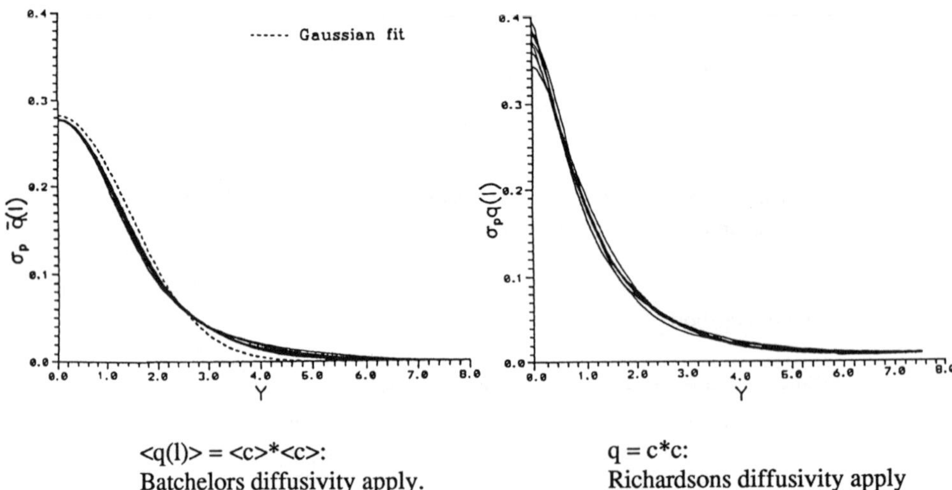

<q(l)> = <c>*<c>:
Batchelors diffusivity apply.

q = c*c:
Richardsons diffusivity apply

Figure 6.7. Recent Lidar measurements (Jørgensen and Mikkelsen, 1993)

6.2.6 Lagrangian Similarity Theory (Batchelor 1964)

Continuous line source, ground level release.

Define: <Z>: Mean vertical plume height

$$\frac{d\bar{Z}}{dt} = c u_* g\left(\frac{z}{L}\right) \tag{6.22}$$

Instantaneous Source: Relative diffusion (Batchelor, 1950)

$$\frac{1}{2}\frac{d\sigma_p^2}{dt} \propto t(\varepsilon\,\sigma_o)^{2/3} \qquad \text{small t limit} \tag{6.23}$$

$$\frac{1}{2}\frac{d\sigma_p^2}{dt} \propto \varepsilon\, t^2 \qquad \text{intermediate t limit} \tag{6.24}$$

$$\frac{1}{2}\frac{d\sigma_p^2}{dt} \propto \overline{v^2}\, t_L \qquad \text{large t limit} \tag{6.25}$$

6.3 Gradient Transport - K Theories

6.3.1 The Diffusion Equation

Conservation principle

$$\frac{dc}{dt} = \frac{\partial c}{\partial t} + u_i\frac{\partial c}{\partial x_i} = 0 \tag{6.26}$$

Reynolds decomposition (eg. c = <c> + c´; <c´> = 0) leads to the **Diffusion equation**

$$\frac{\partial \bar{c}}{\partial t} + \bar{u_i}\frac{\partial \bar{c}}{\partial x_i} = -\frac{\partial}{\partial x_i}\overline{u_i' c'} + sources - sinks \tag{6.27}$$

6.3.2 Gradient Transport: K - closure

In analogy to "exact" molecular diffusion: **First order closure:**

$$\overline{u_i' c'} = -K_i\frac{\partial \bar{c}}{\partial x_i} \tag{6.28}$$

6.3.3 Analytical Solution of the Diffusion Equation

Assume one-dimension (y); K_z and <u> are constants $\neq 0$:

$$\frac{\partial \bar{c}}{\partial t} = \bar{u}\frac{\partial \bar{c}}{\partial x} = \frac{\partial}{\partial z}(K\frac{\partial \bar{c}}{\partial z}) = K\frac{\partial^2 \bar{c}}{\partial z^2} \tag{6.29}$$

Solution

$$\bar{c}(z,t) = \frac{Q}{\sqrt{2\pi}\ \bar{u}\sigma_z} e^{-\frac{1}{2}(\frac{z}{\sigma_z})^2} \tag{6.30}$$

where $\sigma_z^2(t) = 2Kt + \sigma_z^2(0)$, since $\bullet\ d\sigma_z^2/dt \equiv K_z$. Analytical solutions also exists when <u> and K_z are power functions in z.

6.3.4 σ_y and σ_z Classification Schemes

2-D solution to the diffusion equation, constant coefficients, is

$$\bar{c}(y,z,t) = \frac{Q}{2\pi\ \bar{u}\sigma_y\sigma_z} e^{-\frac{1}{2}(\frac{y}{\sigma_y})^2 - \frac{1}{2}(\frac{z}{\sigma_z})^2} \tag{6.31}$$

where σ_y and σ_z are given as functions of the Pasquill-Gifford Turner (PGT) stability schemes.

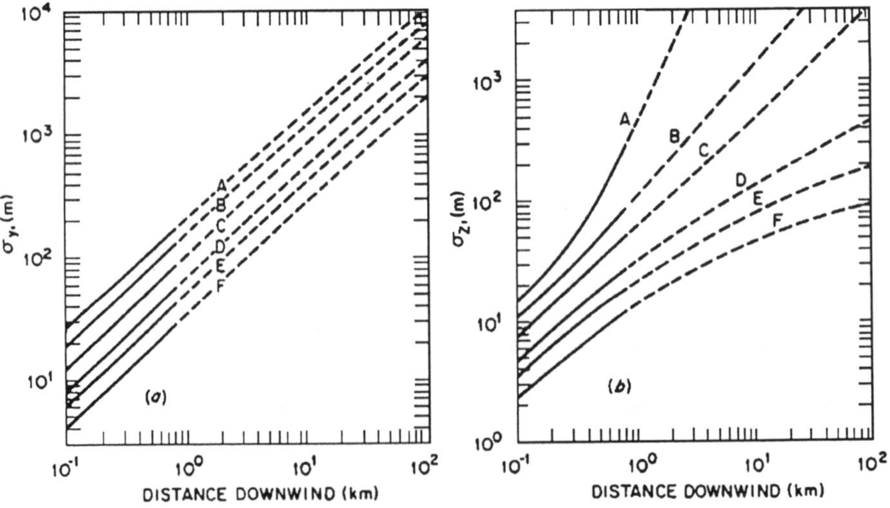

Figure 6.8. Pasquill-Gifford Turner (PGT) stability schemes.

6.3.5 Next Generation of Practical Short-range Dispersion Models (Olesen and Mikkelsen, 1992)

Formula-based, similarity scaling (u_*, L, z_i, w_*):

$$\sigma_y = \sigma_\theta \bar{u}t\, F_y(u_*, w_*, L, z_i, z) \quad (6.32)$$

$$\sigma_z = \sigma_\varphi \bar{u}t\, F_z(u_*, w_*, L, z_i, z) \quad (6.33)$$

Examples from the Danish OML model:

Unstable atmosphere, elevated release: $\sigma_z^2 = 0.33\, w_*^2 t^2 + 1.2\, u_* t^2$

Stable atmosphere, z > L: $T_L = 0.45\, L/u_*$.

6.3.6 Similarity theory for K_z

Surface layer (Businger et al., 1971):

$$K_z = \frac{\kappa u_* z}{\Phi_h(z/L)} \quad (6.34)$$

where κ: von Kàrmàns constant, and Φ_h: dimensionless temperature gradient (for heat).

Boundary Layer (Hanna, 1968)

$$K_z = 0.15\sigma_w \lambda_m \quad (6.35)$$

where

- σ_w: Vertical variance.
- λ_m: Peak wavenumber in w-spectrum (Kaimal et al. 1976).

6.3.7 Spectral Closure of the Diffusion Equations.

Fourier-transforming the diffusion equations (1-D in z) leads to (Troen et. al., 1980):

$$\frac{d}{dt}\tilde{c}(k,t) = -\, k^2 K(k)\tilde{c}(k,t) \quad (6.36)$$

where

$\frac{1}{2}\, d\sigma_z^2/dt = K_z(\sigma(t)) = K_z(\alpha/k)$; [non-local gradient].

6.4 Diffusion Parameterization in Mesoscale Models

6.4.1 A Linearized Model for Terrain-induced Flow (Santabarbara et al. 1993)

$$
\begin{aligned}
U_0\frac{\partial u}{\partial x} + V_0\frac{\partial u}{\partial y} &= -\frac{\partial p}{\partial x} + fv + K\frac{\partial^2 u}{\partial z^2} , \quad \frac{\partial u}{\partial x} + \frac{\partial v}{\partial y} + \frac{\partial w}{\partial z} = 0 , \\
U_0\frac{\partial v}{\partial x} + V_0\frac{\partial v}{\partial y} &= -\frac{\partial p}{\partial y} - fu + K\frac{\partial^2 v}{\partial z^2} , \quad U = U_0(c_1 \mid \alpha \mid^{-1}) , \\
U_0\frac{\partial w}{\partial x} + V_0\frac{\partial w}{\partial y} &= -\frac{\partial p}{\partial z} + K\frac{\partial^2 w}{\partial z^2} , \quad K = K(c_2 \mid \alpha \mid^{-1}) ,
\end{aligned}
$$

Figure 6.9. Linearized Model for Terrain-induced Flow

Restrictions

Near-neutral stratification.
Surface-layer profiles for <u> and K_z.

Boundary Conditions

$\langle u\rangle \cdot \nabla h(x,y) = w(0)$; h(x,y) is terrain.

Closure

Spectral $K(k) = K(\alpha/k)$

Solution

Analytic in z for each wave-number, viz.:

$$\tilde{u}(k,z) = k\tilde{h}(k)\bar{u}(k)\left(e^{-kz} - e^{-\frac{z}{l}(i-1)/\sqrt{2}}\right) \tag{6.37}$$

where

$\tilde{h}$ (k):	Fourier-transformed terrain,
L = 1/k:	Outer length scale (invicid limit),
l(k):	Inner length scale, below which stress-influences.

Summary

+ : Modest computational effort
- : Truncated physics, 1st order perturbations only

6.4.2 Mesoscale - Primitive Equation based Diffusion

Assume

Navier Stokes Equation s.

Boussinesq approximation (ρ only in w: δρ/ρ= g´).

Shallow system ($\partial u_j/\partial x_j=0$)

Small slopes

Hydrostatic assumption

Corresponding conservation equations in terrain following coordinates (Pielke, 1984):

$$\frac{\partial}{\partial \bar{\bar{x}}^j} \rho_0 \frac{s - z_G}{s} \bar{\bar{u}}^j = 0,$$

$$\frac{\partial \bar{\bar{u}}^1}{\partial t} = -\bar{\bar{u}}^j \frac{\partial \bar{\bar{u}}^1}{\partial \bar{\bar{x}}^j} + \left(\frac{s}{s - z_G}\right)^2 \frac{\partial}{\partial \bar{\bar{x}}^3} K_M \frac{\partial \bar{\bar{u}}^1}{\partial \bar{\bar{x}}^3} - \overline{\tilde{u}^{j''} \frac{\partial \tilde{u}^{1''}}{\partial \bar{\bar{x}}^j}}\Bigg|_{j=1,2}$$

$$- \bar{\theta} \frac{\partial \bar{\pi}}{\partial \bar{\bar{x}}^1} + g \frac{\sigma - s}{s} \frac{\partial z_G}{\partial x} - \hat{f} \bar{\bar{u}}^3 + f \bar{\bar{u}}^2,$$

$$\frac{\partial \bar{\bar{u}}^2}{\partial t} = -\bar{\bar{u}}^j \frac{\partial \bar{\bar{u}}^2}{\partial \bar{\bar{x}}^j} + \left(\frac{s}{s - z_G}\right)^2 \frac{\partial}{\partial \bar{\bar{x}}^3} K_M \frac{\partial \bar{\bar{u}}^2}{\partial \bar{\bar{x}}^3} - \overline{\tilde{u}^{j''} \frac{\partial \tilde{u}^{2''}}{\partial \bar{\bar{x}}^j}}\Bigg|_{j=1,2}$$

$$- \bar{\theta} \frac{\partial \bar{\pi}}{\partial \bar{\bar{x}}^2} + g \frac{\sigma - s}{s} \frac{\partial z_G}{\partial y} - f \bar{\bar{u}}^1,$$

$$\frac{\partial \bar{\pi}}{\partial \bar{\bar{x}}^3} = -\frac{g}{\bar{\bar{\theta}}} \frac{s - z_G}{s},$$

$$\frac{\partial \bar{\theta}}{\partial t} = -\bar{\bar{u}}^j \frac{\partial \bar{\theta}}{\partial \bar{\bar{x}}^j} + \left(\frac{s}{s - z_G}\right)^2 \frac{\partial}{\partial \bar{\bar{x}}^3} K_\theta \frac{\partial \bar{\theta}}{\partial \bar{\bar{x}}^3} - \overline{u^{j''} \frac{\partial \theta''}{\partial \bar{\bar{x}}^j}}\Bigg|_{j=1,2} + \bar{\bar{S}}_\theta,$$

$$\frac{\partial \bar{q}_n}{\partial t} = -\bar{\bar{u}}^j \frac{\partial \bar{q}_n}{\partial \bar{\bar{x}}^j} + \left(\frac{s}{s - z_G}\right)^2 \frac{\partial}{\partial \bar{\bar{x}}^3} K_\theta \frac{\partial \bar{q}_n}{\partial \bar{\bar{x}}^3} - \overline{u^{j''} \frac{\partial q_n''}{\partial \bar{\bar{x}}^j}}\Bigg|_{j=1,2} + \bar{\bar{S}}_{q_n},$$

$$\frac{\partial \bar{\chi}_m}{\partial t} = -\bar{\bar{u}}^j \frac{\partial \bar{\chi}_m}{\partial \bar{\bar{x}}^j} + \left(\frac{s}{s - z_G}\right)^2 \frac{\partial}{\partial \bar{\bar{x}}^3} K_\theta \frac{\partial \bar{\chi}_m}{\partial \bar{\bar{x}}^3} - \overline{u^{j''} \frac{\partial \chi_m''}{\partial \bar{\bar{x}}^j}}\Bigg|_{j=1,2} + \bar{\bar{S}}_{\chi m},$$

Figure 6.10 RAMS

6.4.3 Prandtl type Closure

Prandtl (1932):

$$K \approx l^2 \left| \frac{d\bar{u}}{dz} \right| \tag{6.38}$$

Example: Let $d\langle u\rangle/dz = ku_*/z \;\Rightarrow\; K = ku_*z$

Problems occur at $d\langle u\rangle/dz \approx 0$ (no flux).

6.4.4 One-equation models

Prandtl & Kolmogorov (1940's)

$$K \approx c_k l \sqrt{e^2} \tag{6.39}$$

where

- e^2: Turbulent Kinetic Energy (TKE equation required).
- l: length scale (local flow), eg.

Blackadar (1962):

$$l = \frac{\kappa(z - z_o)}{1 + \dfrac{\kappa(z - z_o)}{l_\infty}} \tag{6.40}$$

with l_∞ ~10 - 25 meters.

6.4.5 Two-equations K-ε models

In a K-ε model are "memory effects" introduced in the diffusivity parameter K in order to reflects the upstream history of the flow. This is accomplished by introduction of an additional transport equation for the dissipation quantity itself ε (or alternatively through an equation for its related length scale parameter l_ε, see Launder and Spanding (1974)).

Harlow & Nakayama (1968) proposed that the turbulent diffusivity could be expressed in terms of:

1) The Total Kinetic Energy: e^2 [m^2/s^2], and
2) The TKE's dissipation rate ε: [m^2/s^3], - or its corresponding length scale l_ε, as:

$$K = c_\mu \frac{(e^2)^2}{\varepsilon} = \frac{c_l}{c} e l_\varepsilon \tag{6.41}$$

where

$$\varepsilon = c\frac{e^3}{l_\varepsilon} \tag{6.42}$$

The turbulent kinetic energy is in a K-ε model obtained from the TKE-equation in the usual form:

$$\frac{d\,e^2}{dt} = production + diffusion - \frac{e^3}{l_\varepsilon} \;\; (TKE) \tag{6.43}$$

whereas a second transport equation for ε is added:

$$\frac{d\varepsilon}{dt} = production + diffusion - \frac{\varepsilon^2}{e^2} \;\; (Dissipation\ equation) \tag{6.44}$$

Solution: Equations. (6.43) and (6.44) are solved simultaneously for e and ε. The diffusivity K is then given directly as: $c_\mu\ e^4/\varepsilon$, or alternatively, by use of equation (6.42), in terms of a turbulent velocity and length scale as: $c_1/c\ el_\varepsilon$.

6.4.6 Large Eddy Simulation (LES)

LES is a numerical method to solve the full, time-dependent Navier-Stokes equations (in their constant-density Boussinesq form) for moment, heat and pollutants, based on parameterisation of the sub-grid scale (SGS) turbulence, see Schumann (Ibid).

LES methods resolves the large scale turbulent motion from first principles by numerically solving a "filtered " set of equations governing this large scale, three-dimensional time dependent motion. Diffusion modelling is here employed only to approximate the effects of the "sub-grid" scale turbulence.

Sub-grid scale turbulence (Reynolds-stresses) are in LES assumed to follow the Kolmogorov (1941) isotropic scaling theory (with energy cascade according to the $\varepsilon^{2/3}\ k^{-5/3}$ power-law). The dissipation of the corresponding SGS kinetic energy $e''=½\langle u''^2_i \rangle$ can be parameterized according to equation (6.42) above.

An important difference relative to the K-e modelling principle is that the length scale l here is a fixed quantity and equal to the prescribed grid scale Δ (which is typically of the order of 10-100 meters depending on the computer capacity, CPU-time constraints and the stratification).

The sub-grid scale diffusivities (for momentum, heat and pollutants) are also given as above by equation (6.41), but in terms of the sub-grid kinetic energy e" and the associated fixed length scale l=Δ.

LES resolves, - in principle at least, in this way the true turbulent atmospheric motion over the range of scales limited by the outer boundary conditions and down to the sub-grid scale Δ. Pollutants emitted in this numerical turbulence are advected around explicitly, however, sub-grid pollutant diffusion must also be modelled explicitly: For a continuous (single-particle, ensemble-averaged) release by use of K-diffusivity in Equation (41), and for instantaneously released puffs according to Richardson (1926): $\sigma_p^2 \propto \varepsilon T^3$.

6.4.7 Direct Numerical Simulation (DNS)

This method implies a full untruncated numerical solution of the Navier-Stokes equations without any "closure-technique" or parameterisation implied.

While DNS-methods have proven successful to low-Reynolds number application and study cases within the field of computational fluid dynamic (eg. in the study of vortex-interaction in a low-Reynolds number two-dimensional fluid), application of DNS to fully developed boundary layer turbulence is out of the question because of the multitude of scales involved. In order to model atmospheric turbulence and diffusion in just one cubic Km using DNS, approximately 10^{18} grid points would be required. This is because the smallest length scale in atmospheric turbulence (the Kolmogorov's length scale) is of the order of 1 mm only so that each of the 3 spatial dimensions require about 10^6 grid points. In addition must be considered the corresponding integration time requirements. DMS techniques are therefore beyond any conceivable modelling capability based on known computer platforms for modelling of the detailed flow and diffusion patterns in the atmosphere.

Parameterisation of turbulence and diffusion processes in the atmosphere will consequently always be required.

Relevant general Literature

PIELKE, R. A. 1984 Mesoscale Meteorological Modelling, Academic Press, Orlando, Fla.

HANNA, S. 1982 Turbulent Diffusion: Chimneys and cooling Towers. In: (ED.) E. Plate, Engineering Meteorology, Elsevier Scientific publishing company, Amsterdam-Oxford-New York.

ANDERSON, A., TANNEHILL, J.C. & PLETCHER, R.H. 1984, Computational Fluid Mechanics and Heat Transfer, Hemisphere Publ. Corp. New York.

OLESEN, H.R. & MIKKELSEN, T (Eds.) 1992: Proceedings of the workshop "Objectives for Next Generation of Practical Short-Range Atmospheric Dispersion Models", Risø, Denmark, May 6-8 Danish Center for Atmospheric Research (DCAR), P.O.BOX 358, DK-4000 Roskilde. 262 pp.

References

BATCHELOR, G.K. 1950 The Application of the Similarity Theory of Turbulence to Atmospheric Diffusion, Q. J. R. Meteorol. Soc. Vol 76, pp 133-146.

BATCHELOR, G.K. 1952 Diffusion in a field of homogeneous turbulence II. The relative motion of particles, Proc. Camb. Philos. Soc., Vol 48, pp 345-362.

BLACKADAR, A.K. 1962 The vertical Distribution of Wind and Turbulence Exchange in Neutral Turbulence, J. Geoph. Res., vol 67, pp 3095-3103.

BUSINGER, J.A., J.C. WYNGAARD, Y.IZUMI & E.F. BRADLEY 1971 Flux-profile Relationships in the Atmosphereic Surface Layer. J. Atmos. Sci., Vol 28, pp 181-189.

DURBIN P.A. 1980 A Stochastic Model of Two-particle Dispersion and Concentration Fluctuations in Homogeneous Turbulence. J. Fluid Mech. Vol 100, pp 279-302.

GIFFORD; F.A. 1982 Horizontal diffusion in the atmosphere: A Lagrangian dynamic theory. Atmos. Environ., Vol. 16, pp 505-512.

HANNA, S.R., 1968 A Method of Estimating vertical Eddy Transport in the Planetary Boundary Layer using Characteristics of the vertical velocity spectrum. J. Atmos. Sci., Vol 25, p 1026.

HARLOW; F. H. & Nakayama, P. I. 1968 Transport og Turbulent Kinetic Energy Decay , Los Alamos Scientific Laboratory Report LA-3854, Los Alamos, New Mexico.

JØRGENSEN, H.E. & T. MIKKELSEN 1993 Lidar Measurements of Plume Statistics, Boundary Layer Meteorology Vol 62, pp 361-378.

KAIMAL, J.C. J.C. WYNGAARD, D.A. HAUGEN, O.R. COTE & Y. IZUMI 1976 Turbulence structure in the convective boundary Layer. J. Atmos Sci., Vol 33, pp 2152-2169.

KOLMOGOROV, A.N. 1941 The Local Structure of Turbulence in Compressible Turbulence for very large Raynolds Numbers. C. R. Akad. Nauk SSSR, Vol 30, pp 301-305

LAUNDER, B.E. & SPANDING, D.B. 1974 The Numerical Computation of Turbulent Flows. Comput. Methods Appl. Math. Eng., Vol 3, pp 269-289.

LEGG, B.J. & RAUPACH, M.R. 1982 Markov-chain simulation of particle dispersion in inhomogeneous flows: The mean drift velocity induced by a gradient in Eulerian velocity variance. Boundary Layer Met. Vol.24, 3-13.

LUHAR, A.K. & R.E. BRITTER 1989 A random walk model for dispersion in inhomogeneous turbulence in a convective boundary layer. Atmos Environ., Vol. 23,1911-1924.

MIKKELSEN, T. & R. ECKMAN, 1983 Instantaneous observations of plume dispersion in the surface layer. In: Proc. 14 TH International Technical Meeting on Air Pollution Modelling and its Application. Sep 27-30, (Plenum: NY) 1984.

MIKKELSEN, T., S.E. LARSEN & H.L. PÉCSELI 1988 Spectral Parameterization of large-scale atmospheric diffusion. In: Proc. Sixteenth International Meeting on Air Pollution and its Applications. Ed. H. van DOP, Plenum press New York, 1988

MIKKELSEN, T., S.E. LARSEN & H.L. PÉCSELI 1987 Diffusion of Gaussian Puffs. Q.J.R. Meteorol. Soc. vol 113, pp 81- 105.

PRANDTL, L. 1932 Meteorologische Anwendungen der Strömungslehre. Beitr. Phys. Atmos., vol 19, pp 188-202.

RICHARDSON, L.F. 1926 Atmospheric Diffusion shown on a Distance-neighbour Graph. Roc. Roy. Soc., a 110, pp 709.

SANTABARBARA, J.M., T. MIKKELSEN, R. KAMADA, G. LAY & A.M. SEMPREVIVA 1993 LINCOM Wind Flow Model. Risø report. Available from: Dep. of Meteorology and Wind Energy, Risø National Laboratory, DK-4000 Roskilde, Denmark.

TAYLOR, G.I. 1921, Diffusion by continuous movements. Proc. London Math. Soc., Ser. 2, vol 20, 196-211

TASSONE, C., GRYNING S.E. & M. ROTACH 1993 A random walk model for atmospheric dispersion in the daytime Boundary Layer. In: Proc. 20th International Technical Meeting on Air Pollution Modelling and its Application, Eds. S.E Gryning and M. Milan, Plenum Press, New York, 1994.

THOMSON, D.J. 1987 Criteria for the selection of stochastic models of particle trajectories in turbulent flows. J. Fluid Mech., vol 180, 529-556.

TROEN, I., MIKKELSEN T. & S.E. LARSEN 1980 Note on Spectral Diffusivity. J. Appl. Meteorol. vol 19, pp 609-615.

WILSON.J.D., LEGG, B.J. THOMSON, D.J. 1983 Calculation of particle trajectories in the presence of a gradient in turbulent-velocity variance. Boundary layer Met. Vol 27, 163-169.

Address of the author

Seniorscientist, Ph.D Torben Mikkelsen
Department of Meteorology and Wind Energy
Risø National Laboratory
DK-4000 Roskilde, Denmark
E-mail:met-tomi@risoe.dk

VII Climatology of Cities

Patrice G. Mestayer and Sandrine Anquetin

7.1 Introduction

There is no generally-accepted theory of the climatology of cities, yet. The urban climate is altogether an old subject of investigations, since the book of Howard (1833), and a new one because of this lack of generally-accepted framework: new techniques and new approaches are likely to produce new breakthroughs.

This topic is still considered complex and diverse. Only a few engineering or micro-meteorological rules are general enough to be exported from one city to another. The actual urban climate is strongly dependent on both the city geographical situation and its structure, i.e. on the regional meteorology and its interaction with the human-made fabrics. Therefore our traditional understanding of the urban climate does not appear like a continuous, smooth body of knowledge with quantitative relationships and where all new findings find easily their place, but more like many dispersed studies letting appear separate pieces of phenomenological informations and very few **systematic** studies; with the noteworthy exceptions of the continuous works of Oke and colleagues at Vancouver, who are since the seventies gradually constructing the reference corpus of urban energetics.

It must be realised that a large part of our knowledge comes from incident results of research and observations in the neighbouring fields of meteorology, dispersion over complex terrains, and wind engineering. The applied character of this last research field is the reason why we can found innumerable studies of the pressure fields on isolated or small groups of buildings, but much fewer complete descriptions of the flows around them, even fewer systematic studies of these flows, and practically no study of the flows and dispersion through large arrays of building-like obstacles as can be found in city *quartiers*[1].

To summarise, our present knowledge of urban climatology is based on three types of data:

- wind tunnel and water tank simulations -- most of them describe the pressures on, the flow and dispersion around, or the nano-climatology (the "comfort") between, special ensembles of buildings and constructions, for building and planning purpose; only a few laboratory works include systematic studies with the one-parameter-at-a-time rationale.
- on site measurements -- a very large number of them are single-point or single-station data sets; very few of them include enough data to correctly relate the local microscale climatology to the regional meteorology.
- numerical studies -- the oldest ones concerned simulations of the perturbations induced by large cities on the meso-scale climatology, but were realised with meso-scale models that had not the features allowing to simulate or predict the climatology of the cities themselves in detail; it is only the recent development of non-hydrostatic meso-scale models that allows to perform such studies. Since the eighties, the studies of the urban radiative and energy budget led to the development of computer based mathematical models of the urban energetics allowing systematic simulations. Quite recently, the development of large, fast computers allowed to use computational fluid dynamics (CFD) codes to predict the flow and pressure fields around buildings.

To those three types of data, we must add a very recent handful of airborne and satellite comprehensive observations of the urban fabrics.

[1] This French word represents a part of a city, with a connotation of cultural homogeneity. Below, we propose a definition of a "quartier" for micro-meteorological and micro-climatic researches.

A. Gyr and F-S. Rys (eds.), Diffusion and Transport of Pollutants in Atmospheric Mesoscale Flow Fields, 165–189.

At a regional scale, the most striking, and the most studied, climatologic effect of the presence of a city, is the so-called *heat island*, i.e., the increase by several degrees of the temperature of the atmosphere over the city compared to that of the upwind neighbour rural area. This has been recognised since long by meteorologists, because it really has a direct and visible effect on meteorologists' every-day observations. Less easy to assess and to evaluate are the city effects on the regional physics and microphysics due to the altered fluxes of sensible heat and moisture, and the production of pollutant gases and particles: among them are the increases in cloud coverage, precipitation, and fog. Finally, because most of the city inhabitants are indeed living inside the city itself, it is also of importance to evaluate the city climatic auto-impacts, i.e., the dynamic and thermodynamic effects of the urban structures on the climate inside the cities, at small scale and at the pedestrian level.

In this course, we first survey the observations assessing the heat island, its traditional conceptual description, its causes and extents, and the induced climatic perturbations at the regional scale. Then we see the consequences of the heat island on the dispersion at meso and sub-meso scales, and on its simulation. We see that large problems remain with the structure of the flows close to the city roofs: this leads to focus on the *urban canopy*, and review our knowledge of the in-canopy energy budget and pollutant dispersion. Finally we survey the present trends in climatology and dispersion at sub-mesoscales in the urban atmosphere and conclude with their possible extensions in the near future. Because there is no recent "reference" handbook on the subject, this chapter ends with a rather long list of references, although we tried to limit it to the only most valuables works.

Let's first define the *urban canopy*. The urban ground is covered with natural and artificial materials, buildings, constructions and obstacles. The heat budget as well as the flow structure close to the ground depend largely on the height and density of the constructions, The flow structure varies from series of turbulent wakes due to isolated obstacles to strongly interacting turbulent streaks and even to turbulent flows having some similarities with flows in semi-porous media. By analogy with the atmospheric layer between the ground and the trees of dense forest, the atmospheric layer below the roof level is called the urban sub-layer or urban canopy. Kondo and Yamazawa (1986) proposed to define its height by the integral of the building heights:

$$h = \frac{1}{S} \sum H_i \cdot S_i \qquad (7.1)$$

where S is the total area under consideration, H_i and S_i the height and area of the individual obstacles (buildings).

7.2 The urban heat island

7.2.1 Description. The systematic temperature difference between the atmosphere of the cities and that of the neighbourhoods have been detected in the nineteenth century by European scientists examining historical records of large cities, including Paris, Berlin, Vienna and London (Landsberg, 1956). It rapidly appeared that the temperature differences $\Delta T_{u\text{-}r}$ increase with time passing and are dependent on the city size, from 2°C for small towns up to 12°C downtown very large conurbations. They are also quite variable during the diurnal cycle and dependent on the meteorological conditions with, e.g., maxima appearing by the end of the afternoon or the beginning of the night in conditions of weak winds and clear air, and in winter.

Oke (1990) reviewed about 700 papers on this subject. Several physical processes have been considered to take part in the heat island generation:

- anthropogenic sensible heat production, due to house heating, vehicles and industries;
- reduction of wind speed, due to the presence of buildings;

- alteration of the water balance, due to the reduction of vegetation evapo-transpiration and to the water vapour direct releases by industrial activities;
- alteration of the radiative budget, including the effects of greenhouse gas production;

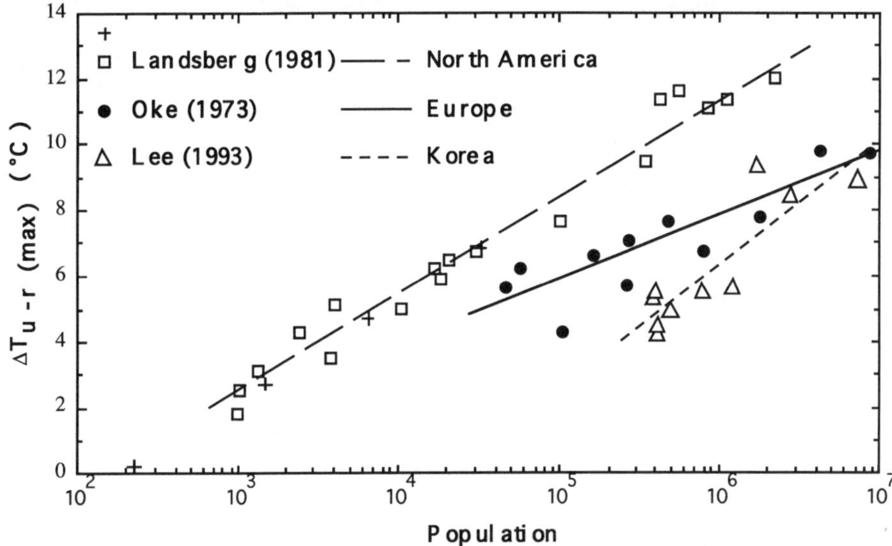

Figure 7.1. Intensity of the heat island, as a function of city size.

- heat storage into the constructions;
- local topography and local wind flows;
- alteration of the diurnal cycle of sensible heat flux, radiative heat flux, and conductive heat flux, especially the reduction of nocturnal cooling.

The role and the importance of these various processes appear to be much dependent on the solar cycle and the season; e.g., the alteration of the water budget is a dominant factor in summer, while the winter dominant factor seems to be the anthropogenic heat flux.

A large number of measurements have been done to assess these effects in all types of cities, but most of the safest conclusions we can draw are issued from the three following large experiments:

- Vancouver, Canada, an experiment mainly realised and analysed by Oke and colleagues (Oke, 1976);
- St. Louis, USA, the METROMEX project from 1974 to 1976 (see references in Changnon et al., 1981, and Changnon, 1992);
- Toulouse, France, an experiment realised in 1978, and analysed mainly by Estournel (1982).

These experiments allowed to evaluate the amplitudes and the variations of the heat island. The magnitude of the heat island is well illustrated by the Figure 7.1. It is a combination of the data reviewed by Landsberg (1981), Oke (1990), and Lee (1993), demonstrating the heat island dependence on the city size.

The first evidence from this figure is that the temperature difference can attain a very large value in the large and dense megalopolis. It is also interesting to note that even a small city of 1000 inhabitants, actually a village, generates a noteworthy heat island. The three lines drawn by Landsberg, Oke, and Lee, through American, European, and Korean city data demonstrate that the structure of the city is also a factor of high importance, probably through the effect of ventilation that depends not only on the building density but also on the building height.

7.2.2 The heat island model. These studies have demonstrated that the heat island effect is important not only close to the ground but also higher in the atmospheric boundary layer and that it generates either the "heat bubbles" or "heat plumes" sketched on Figure 7.2. Due to the city induced thermal and dynamic convection, the frequency and the intensity of the inversion layers tend to decrease over the urban areas and the urban boundary layer therefore tends to be more often close to neutral.

The Figures 7.2 (a) and (b) define the "urban boundary layer" structure in conditions of (a) moderate to high winds, and (b) weak winds. These classical pictures must be seen as a

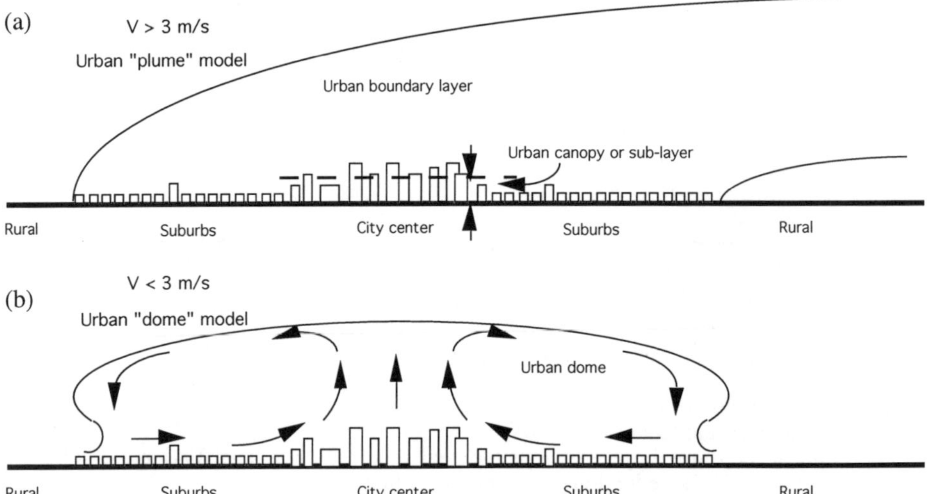

Figure 7.2. Models of the urban heat island in conditions of (a) moderate to strong wind and (b) weak wind.

crude approximation, a simplistic and purely descriptive scheme corresponding to the pioneering modelling efforts of the 1970's. Actually all meteorological conditions produce some combination of these two situations, and the picture is much more complicated: the structure of the "urban boundary layer" is totally 3-D, depends largely on the local orography and on the inhomogeneity of the different quartiers inside the city. Nevertheless it is important to realise that most of the studies of the last two decades have been developed in the framework of these conceptual models.

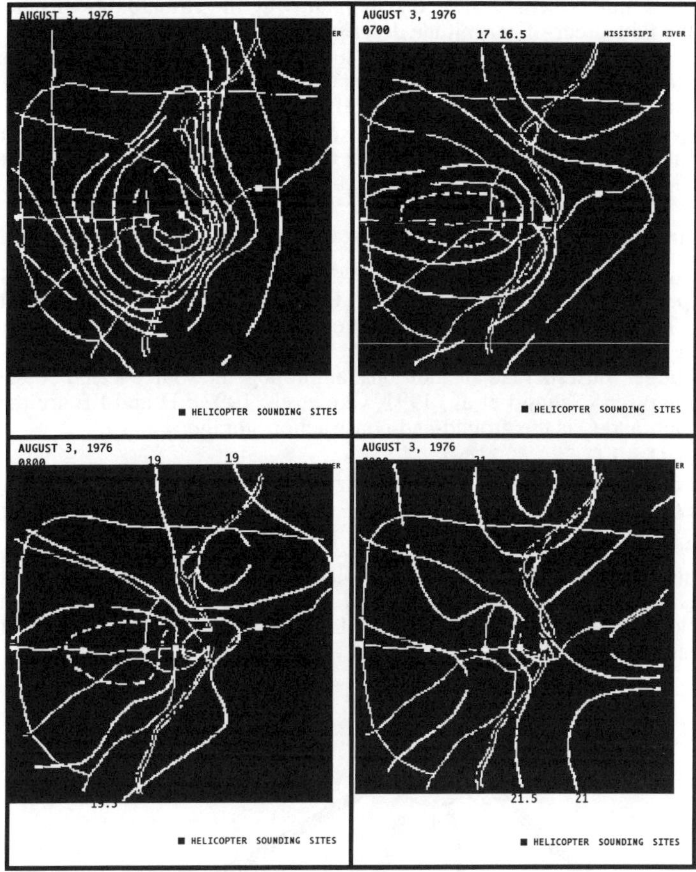

Figure 7.3. Isotherm distributions deduced from land based and helicopter borne measurements over St. Louis, Missouri, during METROMEX, on August 3, 1976, every hour since 6 am (Godowitch, 1986). Note at 6:00 the typical pattern looking like topographic contours of an island: the *heat island*.

The aerodynamic model behind the scheme (a) is that of the development of an internal layer inside a 2-D boundary layer over a flat plate, due either to a roughness change (dynamic internal layer) or to a change in the surface heat flux (thermal internal layer). These model flows have been extensively studied in wind tunnels. The scheme (a) sees the urban boundary layer as one (or a series of) internal layer(s) in development in the atmospheric boundary layer. It therefore sees the city constructions essentially as a carpet of roughness elements and the city landscape as a rough terrain.

Yet, when looking at the actual details of the measurements over large urban areas, it appears that the distribution of the heat island is both very complex in space and rapidly changing in time. For instance the four successive surface temperature distributions displayed in Figure 7.3 (Godowitch et al., 1986) show that the internal patterns of the heat islands are both rather complicated and quite rapidly changing during the diurnal cycle, due to the 3-dimensional influence of the terrain inhomogeneity, although they have been obtained over the relatively flat area of St. Louis, Missouri. The variability and complexity

of the processes hidden behind these data explain why the climatology of cities is still in the state of the first-order-of-magnitude description..
Before going any further we need to analyse the difference in the atmospheric energy budgets over city surfaces and over rural areas to understand the origin of this heat island.

7.2.3 Energy budget. Following Todhunter & Terjung (1988), the steady state energy budget over a flat surface can be written under the following form:

$$(Q + q)(1 - a) + \varepsilon I_{total} - \varepsilon\sigma T^4 = H + G + LE \tag{7.2}$$

where Q = direct solar radiation, q = diffuse solar radiation, a = surface albedo, ε = surface emissivity, I_{total} = hemispherical long wave radiation, σ = Stefan-Boltzmann constant, T = surface temperature, H = sensible heat flux, G = conduction, and LE = latent heat flux (Figure 7.4). The equation (7.2) has been written so as to let appear the radiative budget on the left hand side and the heat budget on the right hand side. Actually, over the urban canopy the budget must include an additional anthropogenic source term F, essentially due to fuel consumption (Schmid et al., 1991, Oke et al., 1992); H and LE are the turbulent fluxes in the air, and G is the ground-and-construction storage:

$$R + F = H + G + LE \tag{7.3}$$

Considering the urban canopy as a nearly horizontal surface, the radiative budget R can be re-written under the simplified form

$$R = (1 - a) \cdot S + \Delta L = S - a \cdot S + L_a - L_s \tag{7.4}$$

where S is the received (direct + diffuse) solar radiation, $a \cdot S$ is the part of the solar radiation that is reflected by the canopy, L_a is the long wave radiation emitted by the atmosphere downwards to the ground and L_s the long wave radiation emitted by the canopy upwards to the sky .

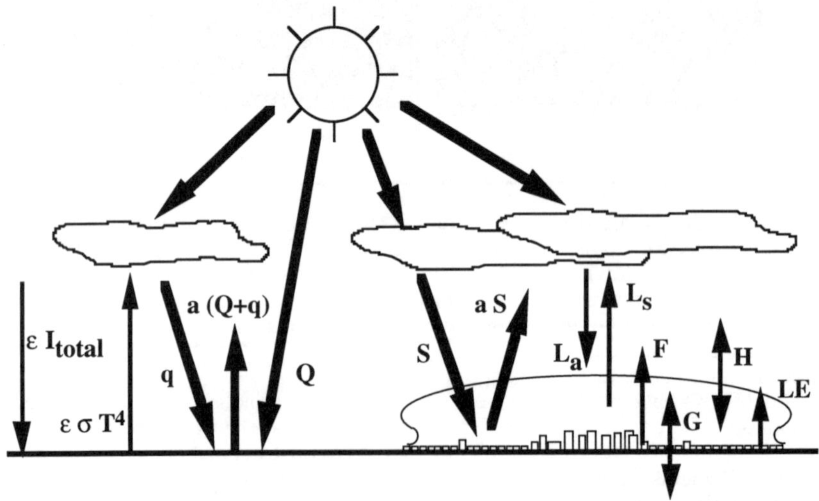

Figure 7.4. Schematic radiative and energy budgets.

The incident solar radiation S can be attenuated by 2 to 20 % over urban areas. The atmospheric pollution is the main cause of this reduction. The short wave (solar) radiations are absorbed by small aerosols, water vapour, oxygen and ozone. The emission of aerosols

also produces condensation nuclei that increase the formation of clouds that also absorb short and long wave radiations.
The part of the solar radiation that is re-emitted to the atmosphere by the urban canopy is essentially depending on the albedo of the urban surfaces. The albedo of a surface characterises the visible radiation reflection of the surface: a surface that perfectly reflects all radiations has an albedo of 1 while an albedo of 0 characterises a surface that totally absorbs all visible radiations. The albedo of a surface depends on its material, on its surface state, on its colour. It can vary with time, due to surface processes as, e.g., the deposition of dust. or the elevation of the sun (glass surfaces have albedos varying from 0.08 to 0.5 as a function of solar incidence angle). The urban surfaces present a very large palette of albedos, from 0.08 for tar and asphalt to 0.5 or 0.9 for white-painted walls, but most of them have lower albedos than the rural surfaces, between 0.1 and 0.2: Oke (1978) estimates the average urban albedo to about 0.15 while Dabbert and Davis (1974) measured a difference of 0.03 between the mean urban albedo (0.13) and that of the rural areas (0.16). As a whole, the canopy presents an even lower effective albedo due to the radiative trap effect of the street canyons and building yards: many incoming radiations undergo multiple reflections on the building walls and the interspersed ground surfaces. The building height and density and the solar elevation influence both the radiative trap effect and the shadowing effect of one building on another (Figure 7.5). Oke (1988) showed that the effective albedo of an array of street canyons can vary by as much as a factor 2 under the only combined effect of the solar incidence and the street geometry.

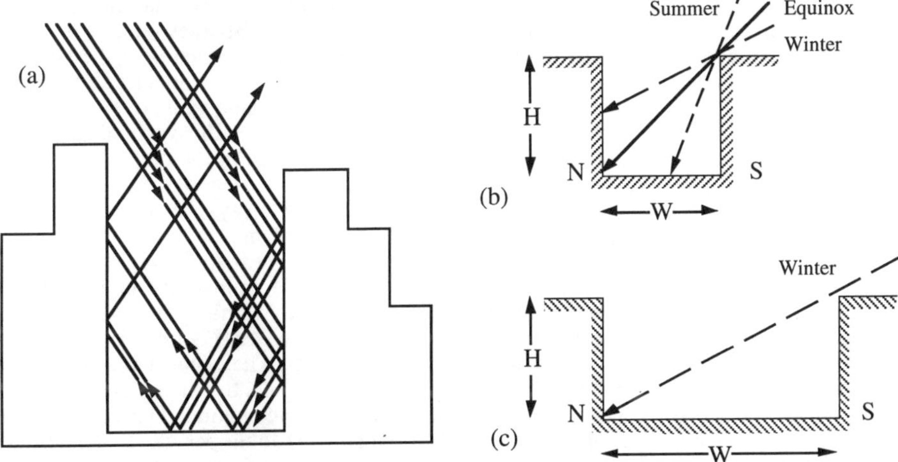

Figure 7.5. The canyon street trapping and shadowing effects on the incident radiations (a after Estournel, 1982, b and c after Oke, 1988)

Due to the presence of more clouds and particles, the long wave radiation emitted downwards by the atmosphere is larger over the city than over the country-side by up to 5 %. The long wave radiations emitted by a surface is proportional to its temperature at the 4th power: the radiative emission of urban grounds and surfaces is considered to be much larger than that of the rural surfaces and it presents a maximum around mid-day. This flux can be as much as 20% higher over cities than over the country-side (Oke, 1979). The lower atmosphere absorption of the long wave radiation emitted by the urban surfaces is also enhanced by the presence of particles and greenhouse gases like carbon dioxide, and the two effects combine to increase the temperature of the urban atmosphere.

The assessment of the various terms in the radiative budget of the urban atmosphere compared to the neighbour rural atmosphere shows that the decrease of solar radiation is quite well balanced by the effect of the lower albedos. The long wave radiations are larger in both directions and these increases tend to balance each other. Finally, although the urban structures do transform, sometimes largely, the various terms of the radiative budget, the net budget imbalances are, in the average, of the same order of magnitude over urban and rural areas. Oke (1979) even consider that there is a slight deficit in the urban radiative budget compared to the rural one. Therefore it does not seem that the radiative budget is the main direct cause of the heat island. So, we must also consider now the heat budget.

The sensible heat flux H is first due to the aerodynamic transfer of heat from the surfaces to the air due to the temperature differences between air and surfaces. The latent heat flux LE is first due to the release of water vapour by the uncovered grounds, the vegetation, and the porous surfaces. But anthropogenic sources of sensible and latent heat are also present in the urban atmosphere: house and factory chimneys, lightning, traffic exhausts. The magnitude of these sources is a function of the city energy consumption, that results of the population density, and also of socio-economical factors as the average standard of life and the city organization. The anthropogenic fluxes increase with the density and with the standard of life; in large, dense, and rich cities of the north, F can be larger than the solar radiative flux. In general H is higher in cities, and LE is smaller. There is less vegetation and less bare soil that are potential surface water storages, but the anthropogenic sources of latent heat can be very intense and are mainly isolated large sources (factory exhausts) that generate strong inhomogeneities into the surface patterns and the atmosphere.

The conduction G is much larger in cities due to the capacity of urban materials to store the heat and to the absence of vegetation. The heat storage varies between 15 and 30 % of the total heat budget. It is also due to the radiative trap effect. The reduction of the vegetal cover and the multiplication of the vertical walls increase the area collecting the solar radiative flux. Although the net radiative budget has about the same value over rural and urban areas, the multiple reflections of the radiations between the vertical surfaces multiply the transfers from radiative energy to heat that is stored in the city fabrics. The heat storage accumulates during the day and is progressively released to the air during the night: conduction G is often the largest sink of heat by day and always the largest source at night, when it almost balances the radiative heat loss. This explains the large values of the heat island during the night and its maximum often observed between 3 and 5 hours after sunset: the country-side stores little heat that can be rapidly re-emitted and cools rapidly after sunset while the heat stored in, and re-emitted by, the city artificial ground and wall materials reduces the cooling of the air included in the canopy layer (Figure 7.6).

Let us also consider the water vapour and latent heat budgets. The water that is stored in the superficial layers of the natural grounds represents also a negative latent heat storage, that is released when the water is released to the air by the vegetation. The evaporation induces a large consumption of heat in the air, about 2 500 Joule per gram of released water (580 Cal/g). In the cities, most of the water that is deposed to the surfaces is collected by the draining network to the sewers and finally to the rivers and the sea, staying in liquid form. Therefore the evaporation is much reduced in the urban atmosphere, and the latent heat that is not pumped from the energy budget is one of the major cause of the heat island.

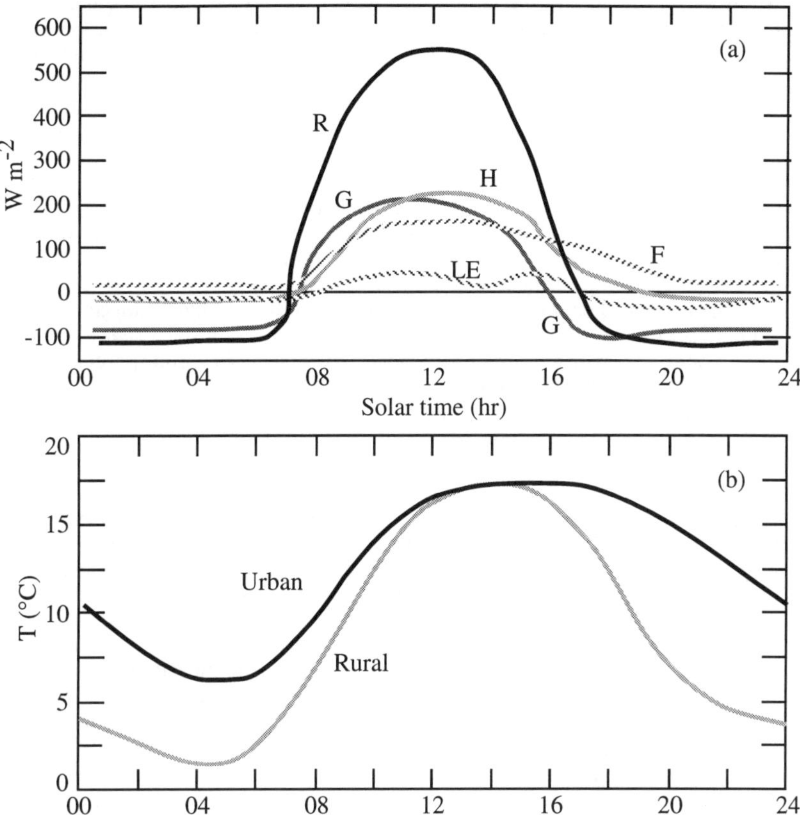

Figure 7.6. Typical, schematic diurnal evolution of (a) the heat budget terms over the urban canopy (symbols as in text) and (b) the urban and rural temperatures (freely adapted from Nunez and Oke, 1976, Oke, 1982, and Oke et al., 1992)

Finally it must be noted that, although a negative temperature difference $\Delta T_{u\text{-}r}$ is rarely observed in city centres, it could well happen quite often in the suburban areas, due to the presence of artificial ground surfaces and the absence of strong heat sources and heat storages (Myrup et al., 1993). This increases the actual complexity of the local flows just over the city and the urban "ground" description for dispersion modelling.

7.2.4 Amplitude, consequences, and modelling. The Figures 7.1 and 7.3 have demonstrated the amplitude and the time variation of the heat island. Other data show that the structure in time and space is actually rather complicated. The heat island appears nevertheless to be a characteristic feature of the urban atmosphere and it can be put clearly in evidence over built-up areas of all sizes, even over small villages or isolated large shopping centres.

The heat island depends on the city size, structure, density, energy consumption, and degree of pollution. Where the pollution is low, the main factor seems to be the impact of solar radiation on the urban surfaces. Where the pollution is high, the energy consumption is also usually quite large and the anthropogenic energy flux seems to be the determinant factor. Of course it is always a combination of the various analysed factors.

The heat island is obviously related to the heat exchanges and energy budget inside the canopy layer, to the exchange at the canopy layer - boundary layer interface, and therefore to the interaction between the meteorology and the canopy structure. This is why several research groups concentrate on modelling the energy budget inside the canopy itself. At least 6 causes of heat build-up can be listed:

- the increase of long wave radiation from the sky,
- the decrease of solar radiation reflection due to the radiative trapping,
- the increase of solar radiation absorption by urban materials due to their lower albedo,
- the increase of heat storage by the urban wall and ground materials,
- the anthropogenic sources of heat and energy,
- the decrease of latent heat consumption due to the reduced vegetation and to the water collection networks.

In the urban boundary layer over the roof levels, (see Figure 7.2), four processes remain as possible causes of the heat island development:

- the warm air provided by the canopy layer,
- the direct anthropogenic heat inputs by the chimneys and high stacks,
- the entrainement of warm air by the induced convective movements,
- the increase radiative transfers due to atmospheric pollution.

Based on a review of experimental data Oke (1973, and followings) proposed to model the heat island effect by expressing the maximum urban-rural temperature difference as a function of the population and of the wind speed as:

$$\Delta T_{u-r} = \frac{P^{0.27}}{4.04 \cdot U_r^{0.56}} \tag{7.5}$$

where P is the number of inhabitants and U_r is the wind speed measured outside the city at a reference level of 10 m. Actually it is obvious when seeing the data of Figure 7.1, that other factors are needed to take into account the city structure. The structure of the heat island is considered to be of the dome type (Figure 7.2b) for reference wind speeds lower than 3 m/s and of plume type (Figure 7.2a) for $U_r > 3$ m/s.

Several parameterisations of the vertical extent of the heat island have been proposed. Among those, Kimura et al's. (1974) relation is:

$$h_{max} = a \cdot R^{-L/6} \qquad \text{with} \qquad R = \beta \cdot g \cdot \Delta T_u \cdot L^4 / \nu_t^2 \tag{7.6}$$

where a is a constant, L is the city length in m, β is the coefficient of thermal expansion per unit volume, g is gravity, ΔT_u is the vertical temperature gradient over the urban area, and ν_t the average turbulent diffusivity (see also Vukovich et al., 1979).

Based on the St Louis METROMEX measurements, Auer and Changnon (1977) proposed to relate the heat island height to the wind speed as:

$$h = 1265 - 87.2\,U \qquad \text{for } 2 \leq U \leq 8 \text{ m/s} \tag{7.7}$$

When the wind is not weak, the heat island is also determined by the development of the dynamic internal boundary layer. Its height is usually parameterized as a function of the fetch x, i.e. the distance to the change of roughness, and of either the roughness lengths of the two surfaces z_{o1} and z_{o2} or that of the rougher surface z_{oM}. Folcher (1989) has summarised the various formulations in the Table 7.1. In addition, Jensen (1981) proposed to take into account the thermal stratification by parameterizing the internal layer height with Monin-Obukhov length scale L as:

$$z_{IBL}(x) = x^{3/2} \cdot L^{-1/2} \qquad \text{for unstable stratification,} \tag{7.8a}$$

$$z_{IBL}(x) = x^{1/2} \cdot L^{1/2} \qquad \text{for stable stratification.} \tag{7.8b}$$

All these parameterisations enter in the framework of the original 2-D scheme of the urban heat island. Some interactions of the heat island with the atmosphere at meso scale, or with the meso-scale climate, have been studied, rarely, with meso-scale meteorological 3-D models, but these have never been used to analyse the 3-D heat island itself at the scale of the city. In the meso-scale numerical models, complex parameterisations of the heat exchanges with bare and vegetated soils have been developed only recently, and they have not been developed for the urban "grounds" (i.e. the canopy) yet. Up to now, the meso-scale simulations considered only simple parameterisations of the canopy "terrain" that appear inadequate particularly with respect to the parameterisation of the surface sensible and latent heat fluxes, the subsurface heat storage, and the role of urban geometry upon surface energy budget (Bornstein & Oke, 1981; Todhunter & Terjung, 1988). Most of our knowledge of the interactions of urban areas with meteorology therefore comes from a small number of experimental results (see also Atwater, 1972, Tatsou Oke, 1981, Mc Elroy, 1972).

Table 7.1 Parameterizations of the internal boundary layer height (after Folcher, 1989)

Authors	Relationships
Elliot (1958)	$z_{IBL}(x) = (0.75 + 0.03 \text{ Log } (z_{o1}/z_{o2}))\ z_{o2}^{1/5}\ x^{4/5}$
Bradley (1968)	$z_{IBL}(x) = x^{4/5}$
Shir (1971), Rao et al. (1974)	$z_{IBL}(x) = x/20$
Bietry et al. (1978)	$z_{IBL}(x) = x/12.5$
Hogstrom et al. (1982)	$z_{IBL}(x) = 0.83\ x^{0.73}$
Wood (1982)	$z_{IBL}(x) = 0.28 \max(z_{o1}, z_{o2})^{1/5}\ x^{4/5}$
Duchene-Marullaz & Sacré (1984)	$z_{IBL}(x) = 0.35 \max(z_{o1}, z_{o2})^{1/5}\ x^{4/5}$
Karlsson (1986)	$z_{IBL}(x) = 1.1\ x^{0.63}$

7.2.5 Meso-scale perturbations induced by urban areas and consequences. The urban "plume" model of Figure 7.2a shows that the climatic perturbations generated by the city are transported by the wind, like a mesoscale "plume" of anthropogenically-modified and polluted air as large as the city itself. This plume can generate a series of perturbations over the downwind regions at distances of several hundreds of kilometres, as, e.g., the "Ozone plumes" during photo-oxydant episodes. This section is essentially a series of quotes of the comprehensive analysis of Stanley E. Changnon (1992) *Inadvertent Weather Modification in Urban Areas: Lessons for Global Climate Change*.

Analysing climatic records, Landsberg (1970) and Huff & Changnon (1973) demonstrated that urban areas influenced warm season rainfall: "the intensive 6-year field studies of urban effects on the St. Louis atmosphere and the 4-year studies at Chicago produced clear evidences that urban influences on convective rainfall (in a humid continental climate) were most evident when the atmosphere was highly unstable and producing heavy (> 2.5 cm) rainfall events (Changnon et al., 1981; Changnon and Semonin, 1979)."

"The urban influences on cloud and precipitation processes have been difficult to measure, and became the focus of several studies in the United States during the 1960s. Changes in clouds and rainfall were suggested in an area extending 50 km east of Chicago (Changnon, 1968), focusing attention on the question of the reality of urban influences on precipitations (Changnon, 1980). Studies of climate records at several North American cities revealed the presence of urban anomalies in precipitation and storm conditions for cities with populations greater than one million (Huff and Changnon, 1973). In general, the area extent and magnitude of the precipitation anomaly were found to be related to the size of the urban area."

"The Metropolitan Meteorological Experiment (METROMEX) studied urban-industrial influences on the atmosphere of the St. Louis metropolitan area to identify the regional influences on clouds and precipitation (Changnon et al., 1981). The urban causes of altered clouds and precipitation included (a) changed in-cloud processes by urban-additions of cloud condensation nuclei; (b) altered fluxes of sensible and latent heat from the metropolitan area, which helped create and enhance vertical air motions; (c) increased convergence and updrafts from momentum shifts due to the city and its structures, and from differential heating, all of which affect the urban boundary layer and perturb the atmosphere; and (d) enhanced moisture, largely from industrial sources (Braham, 1981). METROMEX established the causes of urban-induced cloud changes, the reality and magnitude of warm-season rainfall increases, and the types of increases in storm activity such as more thunderstorms, lightning, hail, and damaging surface winds (Changnon et al., 1981). These changes in cloud and precipitation climate occurred in a broad region embracing the city and a large fan-shaped area extending as far as 65 km east of St. Louis (Changnon, 1979). In a study of summer rainfall over Washington, DC, Harnack and Landsberg (1975) found that the urban area caused showers to develop and concluded that the heat island led to preferential mesoscale forcing and localised cloud development and enhancement."

"The fairly sizeable changes in local, urban, and regional climates due to anthropogenic activities (Changnon, 1973; CEM, 1977) have produced many changes in the biosphere and a mixture of winners and losers in the socio-economic context (Bertness, 1980; Changnon et al., 1979; Davis, 1990; Hoch, 1976; Sassone, 1976). Social attitude assessments revealed that persons living east of St. Louis, where the climate had been altered, either had not detected the climate differences or expressed a lack of concern about the change (Farhar, 1979). In the lee of St. Louis, the 25 % increase in average summer precipitation and reduced visibility due to pollutants have led to more automobile accidents (Sherretz and Farhar, 1978), less leisure time, increased storm damages, more suburban and farm land flooding, and more soil erosion, but a net increase in crop production and increased cropland values (Changnon, 1979b). The migration away from urban centres leading to suburban development over the last 100 years was partly due to the less-than-pleasant climate, including poor air quality, that had developed in the urban centre. Study of the relocation process of wealthier urban residents and the stay-in-place decisions of affected farmers located beyond the cities is important. These findings serve as important lessons for those concerned about the environmental impacts and socio-economic adjustments to larger-scale changed climatic conditions."

7.2.6 Simulation of dispersion over urban areas at meso-scale. The simplest approach consists in considering that the flow over an urban area is similar to the flow over a "normal" rough surface, with a given, large, roughness length z_o and a defined surface heat flux. Assuming that the city is large enough to ensure, locally, a "sufficient" horizontal homogeneity, the Monin-Obukhov parameters can be defined in the usual way and the flow structure can be characterized by the surface stress ρu_*^2 and heat fluxes H_s and LE_s. Due to the large height of the roughness elements, it is needed to introduce a zero-plane displacement height d in the parameterization of the mean wind profile:

$$U(z) = \frac{u_*}{\kappa} \cdot \mathrm{Ln}\left(\frac{z-d}{z_o}\right) \tag{7.9}$$

z_o and d can be either computed from measured profiles of U(z) or deduced from empirical relationships relating these variables to the density and geometrical characteristics of the roughness elements (buildings, trees ...) (see, e.g., Counihan, 1975, Karlsson, 1986, Lettau, 1989) although their applicability to city fabrics seems highly questionable (Rotach, 1991). The meso-scale dispersion calculations based on this approach can be of value when the details of the ground structure cannot influence largely the transport and diffusion, as e.g., buoyant plumes emitted from outside of the city and transported high over the ground in convective meteorological conditions.

A less simplistic approach considers the atmospheric flow transformation due to the internal boundary layer (IBL) in development along the wind direction (2-D model of Figure 7.2). We have seen in the section 2.4 several empirical parameterizations of the IBL depth z_{IBL} as a function of the fetch x. Recently Melas & Kambezidis (1992) reviewed some recent models of the thermal internal boundary layer (TIBL). They divide them in 3 categories: empirical models of the form $z_{IBL} = a\ x^b$ as those of Table 1; similarity models where the growth of the TIBL is parameterized by analogy with the spread of a smoke plume from a ground-level source (an analogy quite valid in the case of near-neutrality for a very weak transition in roughness and surface heat flux, based on the equality of the diffusivities for momentum and matter); slab models assuming that the TIBL is composed of a mixed layer, where the vertical profiles are uniform, bordered by thin layers, surface layer at the bottom, and entrainement layer at the top. In these assumed thin layers, the temperature gradients are forced to match the ground temperature and the unperturbed temperature profile above the TIBL top, identical to the profile upstream of the transition (Figure 7.7)(see also Stull, 1988). Melas & Kambezidis (1992) observe after others, that the models of higher complexity, e.g. Gryning & Batcharova's (1990) slab model, present the best agreements at mesoscale with the experimental data over the Athenian urban coastal area, when meteorological conditions are close to neutrality and the fetch is large.

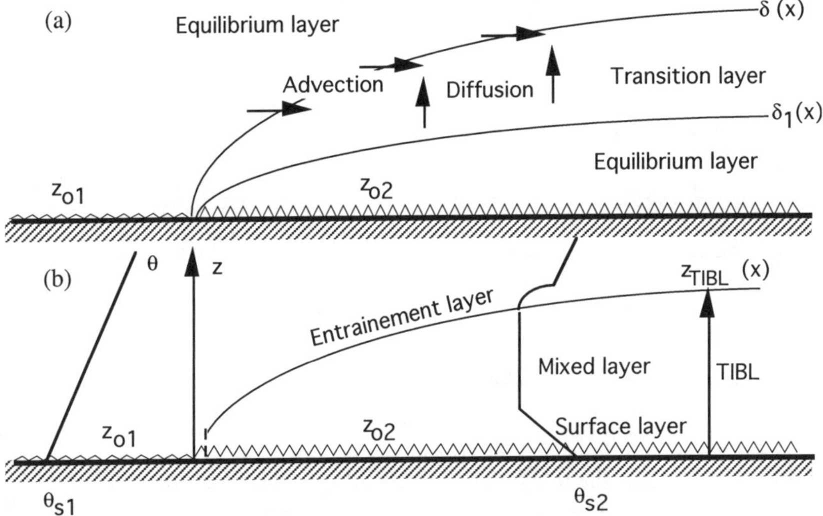

Figure 7.7. Schematic representation of the development of an internal boundary layer at a transition to a rougher surface ($z_{o2} > z_{o1}$) : (a) after Hunt & Simpson (1982) and Folcher (1989) (in the case when $z_{o2} < z_{o1}$, the diffusion is downwards in the transition layer); (b) after Melas and Kambezidis (1992)

It appears that none of these models is able to correctly represent the structure of the flow altogether at short and long fetches (the slab models are better for long fetches, meso-scale dispersion, while similarity models are well adapted to short fetches, small scale dispersion) and to represent the structure of the lowest layers of the urban areas (just over the roofs). Experimental data obtained inside large cities show that the vertical structure of the atmosphere is more complex than that described by Figures 7.2 and 7.7. Most representations put in evidence, over the canopy layer, a perturbed layer where Monin-Obukhov surface layer similarity theory does not hold (Figure 7.8).

Below the roof top height, the urban canopy layer is characterized by flow recirculations and radiative trappings. Just over the roofs, the flow is close to horizontal in the average but highly perturbed by individual building wakes and heat flux variations. Over this transition layer, if the city fabric is not too inhomogeneous, the atmosphere eventually "feels" the lowest layers as a rough warm terrain; its structure can resemble that of a surface layer in (local) inertial equilibrium, and it can be characterized by locally-defined Monin-Obukhov similarity parameters. This "local surface layer" can be capped by a mixed layer that actually is itself a growing TIBL. Simulations of mesoscale dispersions of pollutants generated inside the canopy and diffused up- and out-wards, or of those coming from the city outskirts and eventually penetrating the canopy down to the pedestrian level, must take into account this multiple layer structure of the urban lower atmosphere, that is largely dependent on the structure of the canopy elements.

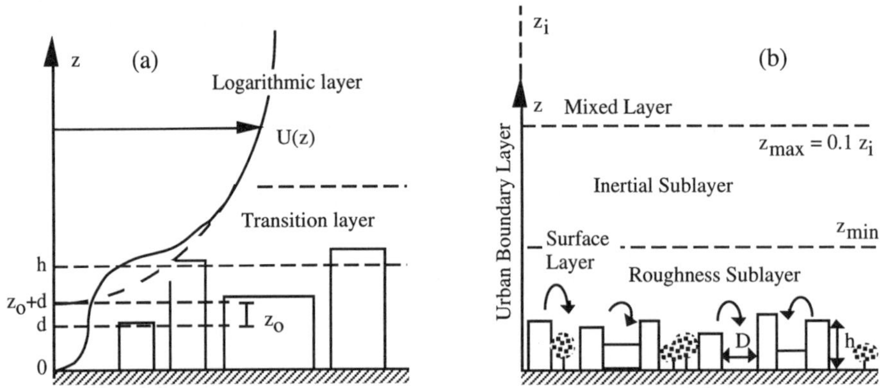

Figure 7.8. Schematic representation of the urban boundary layer : (a) after Karlsson (1986), Folcher (1989); (b) after Oke (1988), Rotach (1991) (z_{min} = 50 - 100 z_o (Tennekes, 1973), = h + 1.5 D (Raupach et al., 1980), = 4.5 h (Garrat, 1978), = 3 D (Garrat, 1980)).

7.3 Climatology and pollutant dispersion in the urban canopy

7.3.1 Street-canyon energy budget. The basic elements of the urban canopy are the isolated or in-group buildings and the street-canyons. The geometry of the buildings, their orientation, their angles, the properties of their surfaces, their albedo and their emissivity, the shadowing effects and the reduction of the optical aperture to the sky, are the physical factors intervening in the radiative budget, and susceptible to modify the energy budget. In addition to a very large number of isolated experiments, a large part of our knowledge comes from the continuous series of studies and numerical simulations of the groups of the University of British Columbia (see, e.g., Nunez and Oke, 1976, Oke, 1973 to 1988) and the University of California Los Angeles (e.g., Terjung and Rourke, 1980; Todhunter and Terjung, 1988). The former used mainly an empirical and experimental approach that allowed to obtain most of the basic insight into the processes taking place into real streets, but difficult to extrapolate to other sites and to build up quantitative predictions. The second group essentially uses and develops numerical models to realise series of academic numerical experiments and process studies, in order to assess the influence of the different climatic and anthropogenic parameters of importance in the energy budget.

The North-South orientation of the street improves the maximum radiative heating of the ground at mid-day. The maximum net radiative imbalance, with clear dry atmosphere, is on the order of 400 W/m^2 at noon. About 60% of the energy is then transformed into sensible heat in the air and 30% is stored in the canyon materials; the remaining 10% are either stored in the ground materials or used for evaporation. Oke (1988) shows that the effective albedo is a combined function of the solar elevation angle and the canyon geometry. He expresses this dependence by means of a sky view factor that is a function of the height-to-width ratio(s) H_i/W where H_i are the building heights and W the street width. An essential issue in the prediction of the energy budget is the effect of the multiple reflections of the incident radiations over several surfaces of the canyon (radiative trap). It goes from a billiard 3-band looking trajectory in the case of moderately high buildings (see, e.g., Noilhan, 1980) to a very large number of successive reflections in the case of arrays of elevated mirror towers. The radiative trap, that is a major cause of heat storage, can also be seen as a reduction factor of the effective albedo (see § 2.3). Here again the two approaches are used to either model the effect by combining theoretical (geometrical) optics and experimental observations (Noilhan, 1980; Nunez and Oke, 1976) or to use numerical codes computing all absorptions and reflections along all possible optical paths of test cases to draw more general simplified relationships (Terjung and Rourke, 1980). It nevertheless appears that the presently available codes produce quite different results when simulating similar test cases. This shows that the present knowledge of the process interactions is insufficient, or that none of these codes incorporate enough of the interacting processes (Todhunter and Terjung, 1988).

7.3.2 Dispersion in the canopy. Up to now, very few studies have combined any two of the three aspects of the dispersion of pollutants inside the canopy layer: dynamic, thermal, and chemical. Most of our knowledge of the in-canopy dispersion comes from simulations in aerodynamic tunnels, essentially in isothermal conditions, very few dynamic measurements on sites, and a handful neutral tracer dispersion experiments at small scales in the atmosphere (see, e.g., in Mestayer, 1991).

As previously mentioned, a relatively large number of studies of the development of internal boundary layers have led to as many expressions of the wind profiles over the canopy, but nearly none for the wind profile inside the canopy layer. One exception is Lettau's (1972) simplified exponential profile

$$U(z) = U_o \cdot \exp(z / z_o) \tag{7.10}$$

where Uo is obtained by matching this profile to the logarithmic profile

$$U(z) = \frac{u_*}{\kappa} \cdot \mathrm{Ln}\left(\frac{z + d + z_0}{z_0}\right) \tag{7.11}$$

at the height of building tops H (see Nicholson, 1975). This expression can be integrated to obtain the mean wind speed <U> in a street parallel to the wind

$$\langle U \rangle = \frac{U_o \cdot z_o}{H} \cdot \left(1 - \exp(H / z_o)\right) \tag{7.12}$$

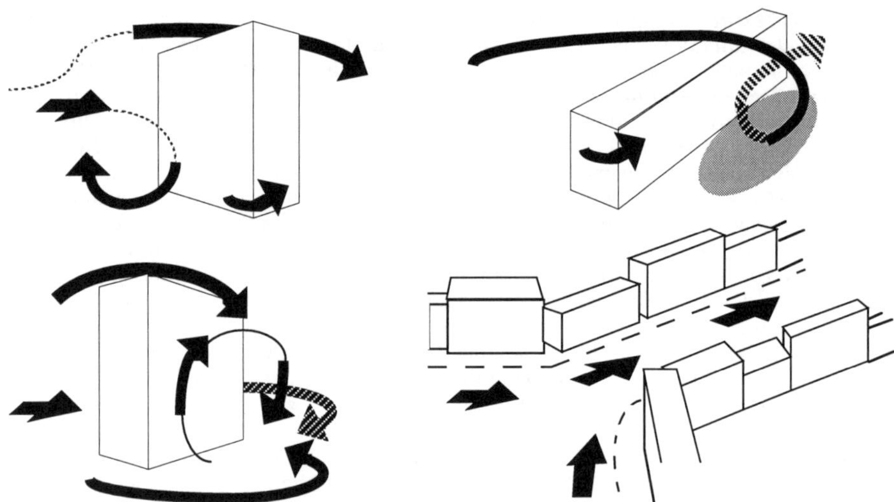

Figure 7.9. Main aerodynamic effects of urban structures (after Gandemer & Guyot, 1976)

Essentially two types of canopy flows have been simulated in wind tunnels: flows around and behind isolated buildings, and in groups of buildings. Most of these studies consist in model simulations of actual or planned constructions and they are not very useful to draw general rules. Also, most of these studies were mainly aiming at describing the effect of the wind on the buildings themselves, the pressure distributions and peaks on the walls, not the flows and the dispersion. These wind tunnel studies can hardly be used to draw general quantitative conclusions although they furnish essential insights of the complexity of the flow structure (Figure 7.9) (Wise, 1971; Wiren, 1975; Gandemer & Guyot, 1976). Therefore only a few tunnel experiments are really useful on the long range.

The flows around isolated block-like buildings have been described by Hosker (1981). It is only relatively recently that numerical simulations of these flows have been possible with full 3-D Navier-Stokes codes including detailed models of turbulence (k-epsilon) (Paterson and Apelt, 1986, 1989, 1990; Murakami et al., 1988). Murakami et al. (1990) validated the k-epsilon numerical simulations versus large eddy simulations and wind tunnel measurements around an isolated cube, but detected some over-predictions of turbulence intensities over the building front edge. As a general rule, the validations over simple building-like geometries show that the "standard" k-epsilon model produces excellent mean velocity fields but tends to slightly over-estimate the turbulent intensities and slightly under-estimate the size of the recirculation zones (e.g., Murakami et al., 1990; Zhang, 1991; Levi Alvares, 1991). These conclusions seem to be true also for the comparisons with on-site measurements around small houses (Bürger et al., 1988; Zhang, 1991).

Wind tunnel experiments of flows around groups of buildings were, and are, mainly aiming at planning pedestrian comfort in order to avoid adverse wind acceleration effects in Venturi-like spaces. They usually show increases of the wind speed between two buildings when the building height increases and when the distances between buildings decrease (Figure 7.9). Ishizaki and Sung (1971) realised systematic studies of arrays of two buildings, while Stathopoulos and Storms (1986) measured the wind into a street-like geometry for various heights, widths, and wind directions. The maximum wind increases, about 40%, are observed for wind directions of 30° off the street axis, quite close to the street entry section; large wind reductions and turbulence intensity increases are observed for angles of 60 and 90° as consequences of the generation of unstable vortices between the two buildings. An increase of the wind speed close to the ground has also been noticed in experiments with buildings of neatly different sizes, due to a return flow from the high pressure on the tall building upwind wall to the low pressures on the small building lee wall. Britter and Hunt (1978) also studied and modelled this configuration. Wiren (1975) made similar systematic wind measurements in L-shaped arrays of two buildings. Hussain and Lee (1980) made a thorough study of the wall pressure distributions on -, and wind profiles between - block-like obstacles of an "infinite" periodic array in a wind tunnel as a function of incident wind characteristics. A group of the University of Cambridge, U.K., is "completing" this experiment with tracer dispersion measurements in similar arrays of "cubes" in a wind tunnel and in an open site, and by "simple" numerical simulations (Hunt and Mullhearn, 1973; Hunt et al., 1991).

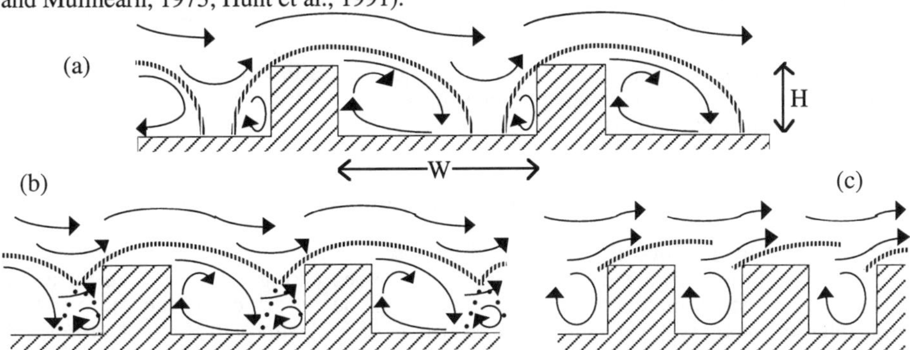

Figure 7.10. Flow regimes over building arrays: (a) Isolated roughness flow; (b) Wake interference flow; (c) Skimming flow.

Oke (1988) analysed Hussain and Lee's data to infer the structure of the flow normal to the street axis in terms of the width-to-height ration H/W: "If the buildings are well apart their flow fields do not interact. At closer spacings ... the wakes are disturbed. When the height, spacing and density of the array combine to disturb the bolster and cavity eddies, this *isolated roughness flow* regime (Figure 7.10a) changes to one referred to as *wake interference flow* (Figure 7.10b). This is characterised by secondary flows in the canyon space where the downward flow of the cavity eddy is reinforced by deflection down the windward face of the next building downstream. At even greater H/W and density, a stable circulatory vortex is established in the canyon and transition to a *skimming flow* regime occurs when the flow does not enter the canyon (Figure 7.10c). The transitions between these three regimes occur at critical combinations of H/W and L/W (where L is the length of the building normal to the flow) ..." Nakamura and Oke (1988) used this analysis to parameterize the in-canyon to over-roof wind ratio. Hunter et al.'s (1992) numerical simulations based on a k-epsilon model show similar flow structures, although their values of the transitions between flow regimes are quite different from those of Oke (1988). Also with k-epsilon numerical simulations, Levi Alvares (1991) demonstrated that the street flow and the in-street scalar dispersion depend not only on the street geometry but also on the wall temperature distribution.

Real size measurements of flows in street are quite rare and we must note the Lagrangian balloon experiment of DePaul and Sheih (1986) in a long street of Chicago, Illinois, Baranger's (1986) measurements in one cross-section of a street of Nantes, France, Nakamura and Oke's (1988) dense documentation of such a cross-section in Kyoto, Japan, and Rotach's (1991) measurements at the canopy-atmosphere interface in Zurich.

There is no model specially developed for simulating contaminant dispersion inside the urban canopy. The only rare relevant and specific models have been developed to take into account the effect of one building wake onto the dispersion of one plume, by extending the gaussian plume approach (Huber, 1984 and other works; Hanna and Paine, 1989) or by simulating the building wake by a secondary source of contaminant (Puttock, 1978; Apsley et al., 1991) (more references can be found in the controversial article of Ramsdell, 1990, and the subsequent discussion of Briggs et al., 1992).
To close this section it must be noted that these studies have been all realised in near neutral conditions, with little or no thermal influence on the flow. The recent numerical simulations realised by the group of Nantes, France (Levi Alvares and Sini, 1992, Sini et al., 1993) clearly demonstrate that the temperatures of the street walls and ground largely influence the in-street flow structures and velocities, and the dispersion of contaminants. Many of the concepts that are now considered as "standards" must be certainly revisited by combining radiative, thermal and dynamic simulations. In addition, a few on-site experiments with a large concentration of sensors allowing extremely detailed documentation of flows and climate in typical canopy geometries are necessary to validate the available numerical models.

7.4 Present researches on the urban atmosphere at meso and sub-mesoscales

7.4.1 In actual large cities, the structure of the lower urban atmosphere and the local air quality are resulting from the interactions of processes taking place at three different typical scales. Most meteorological processes can be described at the meso-scales (atmospheric boundary layer height - several tens of km) although the meteorological forcing is known to be due to larger scales. Many dispersion processes are to be studied in the immediate vicinity of the sources, or of the impact targets, and therefore must be studied at small scales inside the urban canopy; this is especially true for those emissions of pollutants which rapidly transform and chemically interact. Those scales are those of a few buildings or a few streets in interaction (say 20 m in height - 10 to 200 m in width).
But most of the pollutants of interest for evaluating the impacts on human health, vegetation, materials, monuments or on the long term atmospheric content (global change) are secondary compounds, not directly produced by anthropogenic sources. So, when we consider that urban dispersion problems must be now looked most of the time as the dispersion of photo-chemically transformable pollutants, we immediately realize that an intermediate range of processes must be studied. Their scales are those of the turbulent mixing, of the not-so-rapid chemical transformations, and of the interactions between the source layers and the atmospheric boundary layer, because the dispersion processes of photochemical compounds depend on the dynamical, physical, micro-physical, chemical, and radiative structures of the neighbour atmospheric layers. They are the scales of the local convection movements that are specific of the urban environment and due to local, even weak orographies combined with strong inhomogeneities in the surface temperatures and in the canopy dynamical structure (Figure 7.11). We call those scales *sub-meso* because they are smaller than the scales usually simulated by meso-scale models. Indeed the meso-scale models are usually unable to simulate those fine decametric scales and those scales are so inhomogeneous in urban areas that it is not known yet how to model them. Up to now meso-scale models cannot correctly predict the pertinent urban patterns, even combined with sophisticated photo-chemical models, because of their inability to simulate the necessary details of the air movements (Moussiopoulos et al., 1992).

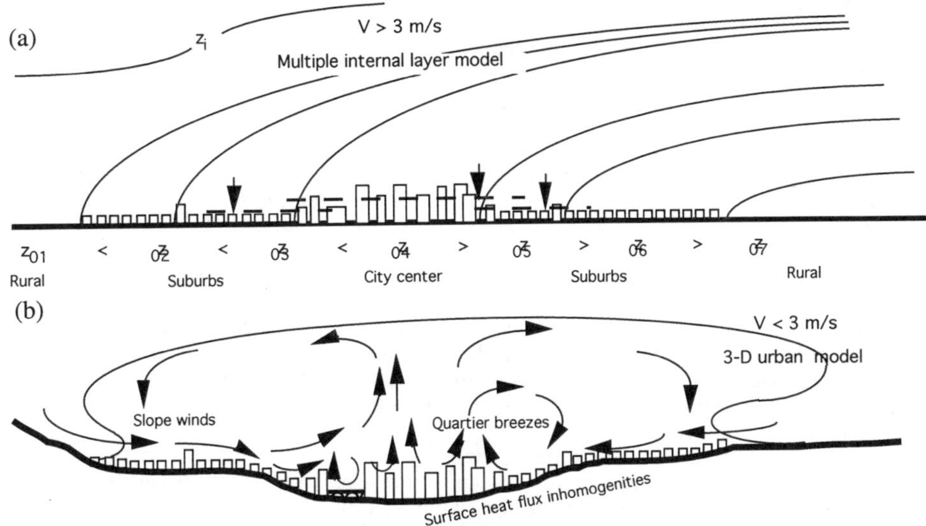

Figure 7.11. Further refinements of the urban atmosphere models.

Urban planners, local decision-makers and city leaders, as well as responsible officials of the environment, need now more complicated data to understand the results of their policy, especially concerning traffic management, to assess the actual air quality with regard to secondary pollutants, to finely predict all-type impacts of new chemical sources, as well as to foresee the effects of new regulations. The need for estimations of secondary pollutant concentrations is especially demanding, since they depend altogether on mixing and reactive processes: this implies a simultaneous knowledge of chemical kinetics and of transport-dispersion behaviour of primary and intermediate compounds.

Under this strong "social demand", the researches are accelerating in this domain. The first effect of this acceleration is a large increase of dispersed and non-coherent works. But this increased social demand takes place at a time of reduced research budgets, which implies a great deal of efforts in rationalizing the researches at national and international levels, especially at the European scale (EC, COST).

7.4.2 At the canopy scale, beside an always-increasing number of in-street measurements combining micro-climatic and air-quality (chemical) parameters, with a large development in Asian countries, and a continuous flow of specific wind tunnel simulations, the recent researches aiming at more general descriptions of the urban climatology develop themselves in the three main following directions (see also Mestayer, 1991).

- 1 - The combination of local on-site measurements with numerical simulations of the radiative heat budget allows to further develop comprehensive urban radiative models and operational codes (e.g., Johnson et al., 1991, Oke et al., 1991, Myrup et al., 1993, Mills, 1993, Mills & Arnfield, 1993)
- 2 - The intensive use of 3-D CFD numerical codes, with high resolution and complete turbulence model, allows systematic studies of in-canopy flows and canopy-atmosphere exchanges, as a function of building geometries and surface temperatures (e.g., Sievers & Zdunkowski, 1986, Hunter et al., 1992, Levi Alvares and Sini, 1992, Sini et al., 1993)
- 3 - The development of street air quality models, that were originally of box-type with primary pollutants, include now some geometrical features of the streets and several

"compressed" chemical reactions to predict secondary pollutants (e.g., Hertel & Berkowicz, 1989)
Up to now these three modelling approaches have not merged yet, but it seems that this should happen soon, especially if co-operative efforts are organized and undertaken.

7.4.3 At mesoscales we observe several efforts to use and/or to adapt mesoscale meteorological codes to the study of the climatology of urban regions (e.g., Bornstein et al., 1986, Chang et al., 1987). Recently, a co-ordinated effort has been undertaken at the European scale, the Athenian Photochemical Smog Intercomparison of Simulations (APSIS). In a series of workshops, meteorological, dispersion, and photochemical codes are applied to data bases documenting events or diurnal cycles over the Athens area (Giovannoni et al., 1993). It appears that several mesoscale models can predict correctly the air flows at the regional scale but that practically none can detail the flows just over roof level responsible for pollutant dispersion in the complex area of Athens, an urban, coastal, hilly, and mountainous site. It also appears that serious tests of the models, especially those combining dynamics and photo-chemistry, require to organize experiments with extensive documentations of (micro-)meteorology, emission inventory and compound monitoring.
Another important mesoscale development is the comprehensive use of satellite observations to analyze the city fabrics and land uses over whole urban areas. The combination of several channels of the same satellite (Lee, 1993) or of different instruments as, e.g., NOAA's AVHRR, Landsat, and SPOT visible and IR channels (e.g., Dousset, 1991), allows detailed analyses of the apparent ground temperature, albedo, reflectance, and heat emission. The pixel sizes can be as small as 20 m (SPOT) and multiple image processings could even provide final resolutions smaller than 10 m. Similar airborne radiative measurements can also be of importance for calibration purpose and finer scale analysis (Takamura, 1992)
As for the simulation of pollutant dispersion at mesoscale it must be noted that, if the urban problems benefit from the development of sophisticated operational models of dispersion over complex terrain, practically none of the operational codes include specific routines for urban terrains. The only exception (to our knowledge) is the Hybrid Plume Dispersion Model (HPDM) which takes into account urban land use categories and their distribution along the wind path as in Figure 7.11a (Hanna and Chang, 1992). Such an urban operational dispersion system (LOSTRAC) is being constructed in co-operation by the Ecole Centrale de Nantes, the Centre Scientifique et Technique du Bâtiment, and Risø National Laboratory. The difficulty in the construction of urban operational codes is that there does not exist reliable-enough parameterizations of the urban grounds' roughnesses and heat fluxes, and of the transformations of the usual orographic effects by the constructions (e.g., downslope "street winds" and "quartier breezes", see Figure 7.11b.)

7.4.4 Sub-meso scales. Because of the large structural, temporal, and spatial variability of the urban lower atmosphere, it appears that the multiplication of measurements will not suffice to assess and to predict the flow structure, the dispersion processes, and the air quality or pollutant contents, and their impacts. The improvement of the knowledge of the combined physics and photo-chemistry of the urban atmosphere calls for the development of a new generation of numerical simulation codes. Indeed partial simulations of the radiative, thermal, and dynamic processes taking place inside the urban canopy are still essential but the assessment of photochemical compound behaviours require combined simulations, including fast photo-chemical reactions. More, impact studies at the scale of the urban area itself requires sub-mesoscale models able to simulate the dynamics and the physics altogether over the whole atmospheric boundary layer height and at a meso-γ scale, and at the decametric scales of the local convective movements in the lower layers that are responsible for the pollutant mixings and the chemical interactions between compounds generated by sources as different as industrial stacks, intense or dispersed traffics, individual heatings, or remote "background pollution".

Not only the detailed prediction, but also the real time control of air quality demand the development of these complex numerical models to understand the relations between the geography, the regional meteorology, the local pollutant sources, and the monitoring measurements. It is one of the prices to pay to obtain the necessary knowledge for the development of the urban operational models of to-morrow.

The development of sub-meso models will allow to study the interactions of the city fabrics and urban sources of heat and pollutants, with the surrounding orography. This is of special importance for the cities located in valleys or valley systems, where the urban radiative budget interacts with the along-valley and katabatic cross-valley winds: this can produce extreme pollutant traps (see, e.g. Stull, 1988).

This model development will be possible only if the structure of the flow in the intermediate layer just over the roofs is much better understood (Figure 7.8). Given the present maximum capacity of the available super-computers (≈ 0.5 Giga word), to simulate an urban area of, say, 10 km x 10 km, with a vertical grid of (at least) 100 levels needed for detailing low level flows, in the first years to come a sub-meso model will not be able to operate over a grid thinner than 20 m x 20 m (or 60 m x 60 m for a 30 km x 30 km domain). Therefore the urban canopy structure will need be replaced by a "terrain" composed of a patchwork of homogeneous quartiers, for which specific *wall laws* will need be defined. These wall laws will be more complicated than the usual roughness length laws based on relation (7.9), as seen in the previous sections. The new parameterizations are also those that are needed to improve operational models: they include not only drag and thermal fluxes but also the canopy ability to exchange matter with the atmosphere downwards and upwards. Also, they will not depend exclusively on the average structure of the quartier but as well on its position with respect to the whole distribution of quartiers and the wind direction. We are actually defining here the *aerodynamic quartier* as a part of a city that has homogeneous features for numerical flow and dispersion simulations.

To develop and assess these parameterizations, three types of researches are urgently needed, preferably in co-ordination:

- 1 - numerical simulations of the development of multiple three-dimensional boundary layers over patchy terrains;
- 2 - numerical simulations of the vertical exchanges in the interfacial layers between canopy and atmosphere, combining dynamic and thermodynamic models in homogeneous quartiers;
- 3 - on-site experiments, documenting with great detail in the same time the flows inside the streets and the flows in the atmospheric layers from the roof level to the inversion height.

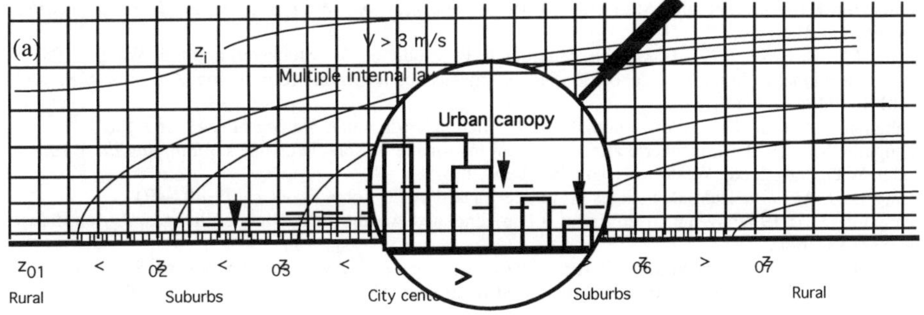

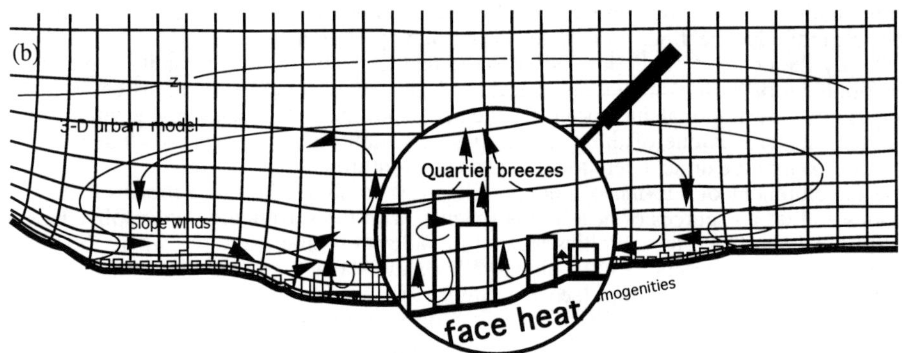

Figure 7.12. Sub-meso research models in development.

Bibliography

Apsley, D.D., A.G. Robins, and J.C.R. Hunt 1991 The BUILD model for predicting building affected dispersion. *EUROMECH 276 Dynamics of the Urban Atmosphere*, 21-24 October, Nantes France, Vol. Proc. p. 34

Atwater 1972 Thermal effects of urbanization and industrialization in the boundary layer. *Boundary-Layer Meteo.* **3**, 229-245

Auer, A.H. and S.A. Changnon 1977 Urban Boundary Layer. *METROMEX* Chap 3

Baranger, P. 1984 Influence des conditions microclimatiques sur la pollution d'une rue. Thèse de 3ème cycle, Université de Nantes.

Bietry J., C. Sacré and E. Simiu 1978 Mean wind profiles and change of terrain roughness. *J. Structural Division* **104**, 1585-1593

Bornstein, R.D. 1986 Application of linked three-dimensional PBL and dispersion models to New York City, *Air Pollution Modelling and its Application V*, D. Wispalaere, edit., pp.543-564.

Bornstein, R.D. and T.R. Oke 1981 Influence of pollution on urban climatology. *Advances in Environmental Science and Engineering* **2**, 171-202

Bradley, E.F. 1968 A micrometeorological study of velocity profiles and surface drag in the region modified by a change in surface roughness. *Quarterly J. Royal Meteo. Soc.* **94**, 361-379

Briggs, G.A., A.H. Huber, W.H. Snyder, R.S. Thompson and J.V. Ramsdell 1992 Discussion on "Diffusion in building wakes for ground-level releases". *Atmospheric Environ.* **26B**, 513-517

Britter, R.E. and J.C.R. Hunt 1979 Velocity measurement and order of magnitude estimates of the flow between two buildings in a simulated atmospheric boundary layer. *J. of Ind. Aerodynamics* **4**, pp 165-182

Bürger, T., J. Franke and G. Tezlaff 1988 Wind measurements in the wake of obstacle. European Community Wind Energy Conference, 6-10 June 1988, Denmark, 170-176

Center for Environment and Man (CEM) 1977 *Report of the Workshop on Inadvertant Weather Modification*, Hatrford, CT, 118 pp.

Chang, J.S., R. Brost, I.S.A. Isaksen, S. Madronich, P. Middleton, W. Stockwell and C.J. Walcek 1987 A three-dimensional Eulerian acid deposition model: physical concepts and formulation, *J. Geophys. Res.* **92**, 14681-14700.

Changnon, S.A. 1968 The LaPorte anomaly-Fact or fiction? *Bull. Amer. Meteor. Soc.* **49**, 165-168

Changnon, S.A. 1973 Atmospheric alterationsfrom man-made biospheric changes, *Modifying the Weather*, Vol. 9, W.D. Sewell, ed., Western Geographic. Serv., 135-184

Changnon, S.A. 1979a Rainfall changes in summer caused by St.-Louis. *Science* **205**, 402-404

Changnon, S.A. 1979b What to do about urban-generated weather and climate changes, *J. Amer. Inst. Planners* **13**, 36-48

Changnon, S.A. 1980 More on the LaPorte anomaly: A review. *Bull. American Met. Soc.* **61**, 702-711.

Changnon, S.A. 1992 Inadvertent Weather Modification in Urban Areas: Lessons for Global Climate Change. *Bull. American Met. Soc.* **73**, 619-627.

Changnon, S.A. and R.G. Semonin 1979 Impact of man upon local and regional weather. *rev. Geophys. Space Phys.* **17**, 1891-1900

Changnon, S.A., D.F. Gatz, J. Betness, S.T. Sonka, J. Bartlett, and J.J. Hasset 1979 Studies of impacts of urban-related weather and climate changes at Chicago and St.-Louis, Contract report 217, Illinois State Watre Survey, 112 pp.

Changnon, S.A., R.G. Semonin, A.H. Auer, R.R. Braham and J.M. Hales 1981 METROMEX: A Review and Summary. *Meteor. Monogr.* No 118, American Met. Soc., 181 pp.

Counihan, J; 1975 Adiabatic atmospheric boundary layers: a review and analysis of data from the period 1880-1972. *Atmos. Environ.* **9**, 871-905.

Davis, R.J. 1990 The legal implications of inadvertant weather modification: METROMEX and the law, Report to NSF Grant ENV 76-80996, Brigham Young Univ., Provo, UT, 38 pp.

DePaul, F.T. and C.M. Sheih 1986 Measurements of wind velocity in a street canyon. *Atmospheric Environ.* **20**, 455-459

Dousset, B. 1991 Surface temperature statistics over Los Angeles : the influence of land use. *Proceedings of IGRASS'91*, IEEE, pp. 367-371.

Duchène-Marullaz, P. 1980 La climatologie urbaine : éléments bibliographiques. Cahiers du C. S. T. B. EN-CLI 80. 9. LDuchene-Marullaz and Sacré 1984,

Elliot, W.P. 1958 The growth of the atmospheric internal boundary layer. *Transaction Americ. Geographic Union* **39**, 1048-1054

Estournel, C. 1982 Etude de l'effet radiatif dans la couche limite : atmosphère urbaine et couche nocturne. Thèse de 3ème cycle, Univ. Paul Sabatier, Toulouse

Farhar, B.C. 1979 The influence of the St.- Louis rain anomaly on human activities, Inst. Behavior Sciences report 16, University of Colorado, 211 pp.

Folcher, A. 1989 Contribution à l'étude de la modification de la structure du vent dans une zone de changement de nature du sol, Thèse de Doctorat, Université de Nantes.

Gandemer J. et A. Guyot 1978 Intégration du phénomène vent dans la conception du milieu bâti. Groupe central des villes nouvelles, Ministère de la qualité de la vie

Giovannoni, J.-M., N. Moussiopoulos and A.G. Russell 1993 Report on the Second APSIS Workshop, 29-30 April 1993, Swiss Federal Institute of Technology, Lausanne (EPFL), Switzerland. *Bull. Amer. Meteorol. Soc.* **74**, 1923-1928.

Godowitch, J.M. 1986 Characteristics of vertical turbulent velocities in the urban boundary layer. *Boundary-Layer Meteo.* **35**, 387-407.

Gryning, S.E. and E. Batchvarova 1990 Analytical model for the growth of the coastal internal boundary layer during onshore flow. *Quart. J. R. Met. Soc.* **116**, 187-203.

Hanna, S.R. and J.C. Chang 1992 Boundary-layer parameterizations for applied dispersion modeling over urban areas. *Boundary-Layer Meteo.* **58**, 229-259.

Hanna, S.R. and R.J. Paine 1989 Hybrid plume dispersion model (HPDM) development and evaluation. *J. Appl. Met.* **28**, 206-224

Harnack, R.P. and H.E. Landsberg 1975 Selected cases of convective precipitation caused by the metropolitan area of Washington, D.C. *J. Appl. Meteor.* **14**, 1050-1060

Hertel, O. & R. Berkowicz 1989 Operational Street Pollution Model (OSPM). Evaluation of the Model on Data from St. Olavs Street in Oslo. *DMU LUFT-A135.*

Hoch, I. 1976 Climate wages and urban scale, *The urban costs of climate modification*, T.A. Ferrar, ed., John Wiley & sons, 171-216

Hogstrom, U., H. Bergstrom and H. Alexandersson 1982 Turbulence characteristics in a near neutral stratified urban atmosphere. *Boundary-Layer Meteo.* **23**, 449-472

Howard, L., 1833 *Climate of London deduced from meteorological observations.* 3rd edn. Harvey & Darton, London.

Huber, A.H. 1984 Evaluation of a method for estimating pollution concentrations downwind of influencing buildings. *Atmospheric Environ.* **18**, 2313-2338

Huff, F.A. and S.A. Changnon 1973 Precipitation modification by major urban areas. *Bull. Amer. Meteor. Soc.* **54**, 1220-1232

Hunt, J.C.R. and P.J. Mullhearn 1973 Turbulent dispersion from sources near two-dimensional obstacles. *J. Fluid Mech.* **61**, 245-274

Hunt, J.C.R., J.C.H. Fung, M.J. Davidson, W.S. Weng, D.J. Carruthers 1991 Modelling flow and dispersion through groups of buildings, *EUROMECH 276 Dynamics of the Urban Atmosphere*, 21-24 October, Nantes France, Vol. Proc. p. 14

Hunter, L.J., G.T. Johnson and I.D. Watson 1992 An investigation of three-dimensional characteristics of flow regimes within the urban canyon. *Atmospheric Environ.* **26B**, 425-432

Hussain, H. and B.E. Lee 1980 An investigation of wind forces on three dimensional roughness elements in a simulated atmospheric boundary layer. Part I: Flow over isolated roughness elements and the influence of upstream fetch, Report BS 55; Part II: Flow over large arrays of identical roughness elements and the effect of frontal and side aspect ratio variations, Report BS 56; Part III: The effect of central model height variations relative to the surrounding roughness arrays, Report BS 57, Dept. Building Sci., Univ. of Sheffield

Ishizaki, H. and I.W. Sung 1971 Influence of adjacent building to wind. *Wind effects on buildings and structures*, Tokyo 1971

Jensen, N.O. 1981 Studies of the atmospheric surface layer during change in surface conditions. *Colloque Construire avec le vent*, CSTB Nantes, Tome I, 4.1-4.20

Johnson, G.T., T.R. Oke, T.J. Lyons, D.G. Steyn, I.D. Watson, and J.A. Voogt 1991 Simulation of surface urban heat islands under 'ideal' conditions at night , Part 1: Theory and tests against field data. *Boundary-Layer Meteo.* **56**, 275-294.

Karlson, S. 1986 The applicability of wind profile formulas to an urban-rural interface site. *Boundary-Layer Meteo.* **34**, 333-355

Kondo, J. and H. Yamazawa 1986 Aerodynamic roughness over an inhomogeneous ground surface. *Boundary-Layer Meteo.* **35**, 331-348

Landsberg, H.E. 1956 The climate of towns. *Man's rolein changing the face of the earth*, University of Chicago Press, 584-603

Landsberg, H.E. 1970 Man-made climatic changes. *Sciences* **170**, 1265-1274

Landsberg, H.E. 1981 The urban climate. *Int. Geophys. Series* **28**, 275 pp.

Lee, H.-Y. 1993 An application of NOAA AVHRR thermal data to the study of urban heat islands. *Atmospheric Environ.* **27B**, 1-13.

Lettau, H. 1969 Note on aerodynamic roughness-parameter estimation on the basis of roughness-element description. *J. Applied Met.* **8**, 828 - 832.

Lévi Alvarès, S. 1991 Simulation numérique des écoulements urbains à l'échelle d'une rue à l'aide d'un modèle k-ε., Thèse de Doctorat, Université de Nantes et Ecole Centrale de Nantes

Lévi Alvarès, S. and J.F. Sini 1992 Simulation of diffusion within an urban street canyon, *First International Symposium on Computational Wind Engineering*, Tokyo, 21-24 Août 1992. *J. Wind Engineering* **82**, 114-119

Mc Elroy, J.L. 1972 A numerical study of the nocturnal heat island over a medium-sized mid-latitude city. *Boundary-Layer Meteorol.* **3**, 442-453

Melas, D. and H.D. Kambezidis 1992 The depth of the internal boundary layer over an urban area under sea-breeze conditions. *Boundary-Layer Meteo.* **61**, 247-264.

Mestayer, P.G. 1991 Editor, EUROMECH 276 Dynamics of the Urban Atmosphere, 21-24 October, Nantes France, Vol. Proc., 37 p.

Mills, G.M. 1993 Simulation of the energy budget of an urban canyon - I. Model structure and sensitivity test. *Atmospheric Environ.* **27B**, 157-170.

Mills, G.M. and A.J. Arnfield 1993 Simulation of the energy budget of an urban canyon - I. Comparison of model results with measurements. *Atmospheric Environ.* **27B**, 171-181.

Moussiopoulos, N., Th. Flassak and Ch. Kessler 1992 Modelling of photosmog formation in Athens, *Air pollution modelling and its applications IX*, A. VanDop and G. Kallos, edit., NATO 17, Plenum Press.

Murakami, S., A. Mochida, and K. Hibi 1988 3-D numerical simulation of air-flow around a cubic model by means of the k-ε model. *J. Ind. Aerodyn.* **31**, 283-303

Murakami, S., A. Mochida, and Y. Hayashi 1990 Examining the k-ε model by means of a wind tunnel test and large-eddy simulation of the turbulence structure around a cube. *J. Ind. Aerodyn.* **35**, 87-100

Myrup, L.O., C.E. McGinn and R.G. Flocchini 1993 An analysis of microclimatic variation in a suburban environment. *Atmospheric Environ.* **27B**, 129-156.

Nakamura, Y. and T.R. Oke 1988 Wind, temperature and stability conditions in an E-W oriented canyon, , *Atmospheric Environ.* **22**, 2691-2700

Nicholson, S.E. 1975 A pollution model for street-level air. *Atmos. Environ.* **9**, 19-31.

Noilhan, J. 1980 Contribution à l'étude du microclimat au voisinage d'un bâtiment. Thèse de 3ème cycle, Université de Nantes

Nunez, M. and T.R. Oke 1976 The energy balance of an urban canyon *J. Applied Meteo.* **16**, pp 11-19

Oke, T.R. 1973 City size and the heat island, *Atmospheric Environ.* **7**, 769-779

Oke, T.R. 1976 Inadvertant modification of the city atmosphere and the prospects for planned urban climates, Proc. *WMO Symposium on Meteorology as related to urban and land-use planning*, Asheville NC, USA, 3-7 Nov. 1975, Note WMO n° 444, 150-175.

Oke, T.R. 1978 *Boundary Layer Climates*, Methuen and Co., Ltd, U.K.

Oke, T.R. 1979 Review of urban climatology 1973-1976, World Meteo. Organ. Technical Note N° 539.

Oke, T.R. 1981 Canyon geometry and the nocturnal heat island: Comparison of scale model and field observations, *J. Climatol.* **1**, 237-254

Oke, T.R. 1982 The energetic basis of the urban heat island. *Quart. J. R. Met. Soc.* **108,** 1-24.

Oke, T.R. 1988 Street design and urban canopy layer climate. *Energy and Building* **11**, pp 103-113

Oke, T.R. 1990 Bibliography of urban climate 1981-1988, WCAP-15, WMO/TD-No. 397, World Meteorological Organization, 62 pp.

Oke, T.R., G.T Johnson., D.G. Steyn, and I.D. Watson 1991 Simulation of surface urban heat islands under 'ideal' conditions at night , Part 2: Diagnosis of causation, *Boundary-Layer Meteo.* **56**, 339-358.

Paterson, D.A. and C.J. Apelt 1986 Computation of wind flows over three-dimensional buildings, *J. Wind Engng. Ind. Aerodyn.* **24** 192-213

Paterson, D.A. and C.J. Apelt 1989 Simulation of wind flows around three-dimensional buildings, *Bldng. Envir.* **24**, 39-50

Paterson, D.A. and C.J. Apelt 1990 Simulation of flow past a cube in a turbulent boundary layer, *J. Int. Aerodyn.* **35**, 149-176

Puttock, J.S. 1978 Modelling the effects of wakes behind hills and buildings on pollutant dispersion, Proc. 9th. NATO-CCMS Int. Tech. Meeting on Air Pollution Modelling and its Application, Toronto, Canada

Ramsdell Jr., J.V. 1990 Diffusion in building wakes for ground-level releases, *Atmospheric Environ.* **24B**, 377-388

Rao, K.S., J.C. Wyngaard and O.R. Coté 1974 The structure of a two-dimensional internal boundary layer over a sudden change of surface roughness, *J. Atmos. Sci.*, **31**, 738-746

Rotach, M. W. 1991 Turbulence within and above an urban canopy, ETH Diss. 9439, publ. as ZGS 45, vdf, Zürich, 245 pp

Rotach, M.W. 1993 Turbulence close to a rough urban surface, Part 1: Reynolds stress, *Boundary-Layer Meteo.* **65**, 1-28.

Sassone, P.G. 1976 Climate modification and some public sector considerations, *The urban costs of climate modifications*, T.A. Ferrar, ed., J. Wiley & Sons, 217-238

Schmid, H.P., H.A. Cleugh, C.S.B. Grimmond and T.R. Oke 1991 Spatial variability of energy fluxes in suburban terrain, *Boundary-Layer Meteo.* **54**, 249-276.

Sherretz, L.A. and B.C. Farhar 1978 An analysis of the relationship between rainfall and the occurence of traffic accidents, *J.Appl. Meteor.* **17**, 711-715

Shir, C.C. 1971 A numerical computation of air flow over a sudden change in surface roughness, *J. Atmos. Sci.* **29**, 304-310

Sievers, U. and W.G. Zdunkowski 1986 A microscale urban climate model. *Beitr. Phys. Atmosph.* **59**, 13-40.

Sini, J.-F., S. Anquetin and P.G. Mestayer 1993 Simulation of diffusion within a heated urban street canyon, *Atmos. Environ.* (to be submitted)

Stull, R.B. 1988 *An introduction to boundary layer meteorology,* Kluwer Acad. Publ..

Takamura, T. 1992 Spectral reflectance in an urban area, A case study for Tokyo, *Boundary-Layer Meteo.* **59**, 67-82.

Tatsou Oka 1980 Thermal environment in urban areas, Swedish Council for Building Research

Terjung, W.H. and P.A. O'Rourke 1980 Influences of physical structures on urban energy budgets, *Boundary-Layer Meteo.* **19**, 421-439

Todhunter, P.E. and W.H. Terjung 1988 Intercomparison of three urban climate models. *Bound. Layer Meteo.* **42**, pp 181-205

Wiren, B. 1975 A wind tunnel study of wind velocity in passages between and through buildings, *Proc. Fourth Int. Conf. on Wind Effects on Buildings and Structures*, Heathrow.

Wise, A.F.E. 1971 Effects due to groups of buildings, *Phil. Trans. Roy. Soc.* **269**, London

Wood, D.H. 1982 Internal boundary layer growth following a step change in surface roughness, *Boundary-Layer Meteo.* **22**, 241-244

Zhang, C.X. 1991 Simulation numérique d'écoulements turbulents autour d'un obstacle, Thèse de Doctorat, Université de Nantes et Ecole Centrale de Nantes

Dr. Patrice G. Mestayer
Eq. Dynamique de l'Atmosphère Habitée
Laboratoire de Mécanique des Fluides
Ecole Centrale de Nantes
44072 NANTES, Cedex 03, France
patrice.mestayer@ec-nantes.fr

Dr. Sandrine Anquetin
Laboratoire des Ecoulements
Géophysiques et Indusriels
Univ. Joseph Fourier & I.N.P.G.
BP 53 X, 38041 Grenoble, France
anquetin@img.fr

VIII Inversion layers

Evgeni Fedorovich

8.1. Introduction

Dealing with diffusion and transport processes in atmospheric mesoscale flow fields one often comes across the situations when these processes take place within so-called atmospheric inversion layers, namely the layers characterized by the increase of the absolute temperature with height. In the course of the diurnal evolution of the atmospheric planetary boundary layer the two most typical examples of the sublayers with the inverse temperature gradient can be observed.

The first one is the stable layer capping the convectively mixed layer (ML) which develops over the heated underlying surface at the daytime and is driven by the positive buoyancy flux from the ground. Strictly speaking, the capping layer is not always strong enough to be classified as a temperature inversion. Still, it is commonly called the capping inversion layer (IL), or capping inversion. Stably stratified inversion layer acts as a lid to the buoyant thermals rising within the mixed layer. The process of interaction between the thermals and the capping inversion is characterized by the entrainment, or mixing down, the less turbulent air from above into the growing mixed layer. That is why the inversion layer in this case is called also the entrainment zone, regardless that the region with inverse absolute temperature gradients may occupy only the part of the layer where entrainment takes place.

The second example is the stable boundary layer (SBL), also called the nocturnal boundary layer, which usually originates at night, where the buoyancy flux at the surface is negative. Like in the capping inversion, in the nocturnal boundary layer vertical transport and diffusion are suppressed by the buoyancy forces. Therefore the turbulence in the stable boundary layer is usually weak and sporadic. The top of SBL is not very well-pronounced and can be defined in most cases merely as the height where turbulence intensity is a small fraction of the surface value. Sometimes SBL can also form during the day, for example when the warm air advection over a cool surface occurs.

Both inversion layers, the one which is capping the convectively mixed layer (elevated inversion), and the surface-based stable (nocturnal) layer, will be subjected to our consideration. We shall briefly discuss the phenomenology of these layers (for details see Stull, 1988), and present model approaches to their description.

8.2. Capping inversion

8.2.1. Characteristics and structure. A convective boundary layer driven by temperature and moisture fluxes at the ground commonly develops during the daytime over land surface. Its growth occurs on the background of stable stratification. Between the well-mixed layer adjacent to the surface and the quiescent layer, there is the capping inversion with strong stable density stratification (Figure 8.1). The inversion (entrainment) layer can be quite thick - averaging about 40% of the depth of the mixed layer. Buoyant thermals rising within the mixed layer overshoot into the capping stable air before sinking back into the ML, and their height of maximum rise defines the top of the entrainment zone. The bottom of entrainment layer is less well defined because there is no sharp demarcation (Nelson et al., 1989). Usually it is taken as that altitude where some small fraction of the air in the horizontal plane has free-atmosphere air characteristics.

A. Gyr and F-S. Rys (eds.), Diffusion and Transport of Pollutants in Atmospheric Mesoscale Flow Fields, 191–211.

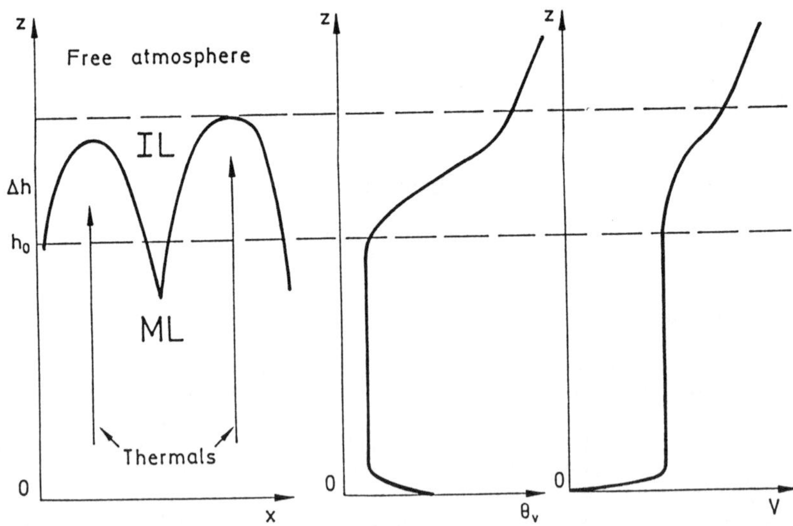

Figure 8.1. Schematic pattern of the capping inversion layer structure. After Stull (1988).

Sharp temperature gradients in the inversion layer are accompanied by substantial changes with height of other meteorological variables, for instance, wind speed and moisture. Pollutants emitted from the sources located within the ML are not able to penetrate into the free atmosphere due to the resistance caused by the static stability in the inversion layer. Trapping of pollutants below the elevated inversion is rather common feature of the convective boundary layer.

The entrainment zone essentially consists of turbulent thermals imbedded within non-turbulent free-atmosphere air. In addition to the large-scale motions associated with the thermals a variety of smaller scale motions exists in the entrainment layer. The latter ones only to small extent contribute to the ML air dispersion from thermals into the free atmosphere.

Kelvin-Helmholtz waves generated in the stably stratified environment of inversion is an important mechanism influencing the entrainment. On the largest scale these waves have the wavelength of the order inversion layer depth. They break, form turbulent spots, and thus contribute to the entrainment. On the smaller scales Kelvin-Helmholtz waves can originate along the top boundary of the thermal. Such short waves appear and decay into turbulence within few minutes. They stipulate the erosion of thermals, but their contribution to overall entrainment is rather small.

Penetration of thermals into the stably stratified air causes also the excitation of gravity waves in the free atmosphere aloft. Propagating at different angles away from the thermals these waves drain kinetic energy from the inversion layer.

In many instances the buoyancy production of turbulent kinetic energy in convective boundary layer considerably dominates over its production due to velocity shear, so the layer can be taken as shear-free. We shall consider one parameterized model for the capping inversion structure corresponding to this case, recently developed by the author in co-operation with Dr. Dmitrii Mironov from ICSC - World Laboratory, Erice, Italy.

8.2.2. Modelling the capping inversion structure. Physical quantity named *buoyancy* can be introduced for characterising the dense difference between the air particle and the surrounding air, and hence the ability of an air particle to rise. Buoyancy is defined as $b = g(\rho_0 - \rho)/\rho_0$, where ρ is the density, ρ_0 is the reference density, and g is the acceleration due to gravity; or, in terms of the virtual potential temperature, θ_v, as $b = g(\theta_v - \theta_{v0})/\theta_{v0}$, where θ_{v0} is the reference value of θ_v. The air of rising thermals is warmer and less dense than the ambient ML air and therefore thermals possess positive buoyancy.

When horizontal averages of buoyancy are calculated from aircraft or lidar measurements data, the resulting buoyancy profile exhibits a smooth behaviour within the inversion layer (Figure 8.2). In the lower part of inversion, the mean buoyancy departs from the value characteristic of the mixed layer (where it is nearly uniform with height), higher on buoyancy sharply increases, reaching its maximum gradient at the height usually close to the middle of the layer, and then matches with the free-atmosphere buoyancy profile.

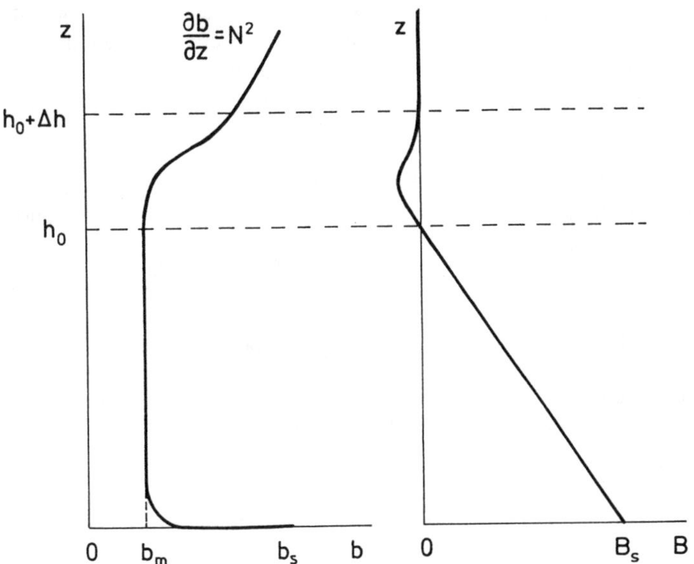

Figure 8.2. Vertical profiles of buoyancy and turbulent buoyancy flux in the convective boundary layer.

The vertical turbulent buoyancy flux decreases linearly with height in the main part of the mixed layer. Its zero crossing level roughly coincides with the bottom of the inversion layer. Being negative all over this layer, buoyancy flux reaches its minimum value within the entrainment zone and vanishes towards its upper boundary. Hence the alternative definition of the entrainment zone is that region where the buoyancy flux is negative.

We simplify the discussion by considering a horizontally homogeneous boundary layer without large scale subsidence, assuming that with the development of convection the vertical buoyancy profile $b(z,t)$ keeps the following form:

$$b = \begin{cases} b_m & \text{at} \quad 0 \le z \le h_0 \ , \\ b_m + \Delta b F(\zeta, G) & \text{at} \quad h_0 \le z \le h_0 + \Delta h \ , \\ b_m + \Delta b + N^2 (z - h_0 - \Delta h) & \text{at} \quad h_0 + \Delta h \le z \ . \end{cases} \tag{8.1}$$

Here $b_m(t)$ is the ML value of buoyancy, $h_0(t)$ is the ML depth, $\Delta h(t)$ is the depth of the inversion layer (IL), $\Delta b(t)$ is the buoyancy increment across IL, $N = \sqrt{\frac{\partial b}{\partial z}}$ is the buoyancy frequency in the free atmosphere, F is function of dimensionless co-ordinate $\zeta = (z - h_0) / \Delta h$ and stratification parameter $G = N^2 \Delta h / \Delta b$.

To provide for the continuity of the buoyancy profile at the boundaries of the IL we impose the following conditions on $F(\zeta, G)$:

$$F\big|_{\zeta=0} = \frac{\partial F}{\partial z}\Big|_{z=0} = 0\,, \quad F\big|_{\zeta=1} = 1, \quad \frac{\partial F}{\partial z}\Big|_{\zeta=1} = G\ . \tag{8.2}$$

The evolution of the buoyancy profile (8.1) should satisfy the buoyancy transfer equation

$$\frac{\partial b}{\partial t} = -\frac{\partial B}{\partial z}\,, \tag{8.3}$$

where B is the vertical turbulent flux of buoyancy.

Integrating (8.3) over z from 0 to h_0 with due regard to the representation (8.1), and defining the top of the mixed layer as the buoyancy flux crossover height , i.e. taking B=0 at $z=h_0$, we obtain the mixed layer buoyancy budget equation

$$\frac{d}{dt}\left[N^2(h_0+\Delta h) - \Delta b\right] = \frac{B_s}{h_0}\,, \tag{8.4}$$

where B_s is the near-surface value of the buoyancy flux.

Integration of (8.3) over z from 0 to $h_0+\Delta h$ gives the equation of total buoyancy budget:

$$\frac{d}{dt}\left\{\frac{1}{2}N^2(h_0+\Delta h)^2 - \Delta b\left[h_0 + (1-C_b)\Delta h\right]\right\} = B_s\,, \tag{8.5}$$

where $C_b(G) = \int_0^1 F(\zeta, G)d\zeta$ is the integral shape factor (Deardorff, 1979).

In accordance with (8.1) and (8.3), in the interfacial (entrainment) layer, at $h_0 \le z \le h_0+\Delta h$, the buoyancy flux profile has the form

$$\begin{aligned} B = &\left[\int_0^\zeta F(\zeta',G)d\zeta' - G\frac{\partial}{\partial G}\int_0^\zeta F(\zeta',G)d\zeta' - \zeta\right]\frac{\Delta h}{h_0}B_s + \\ &+\left[F - G\int_0^\zeta F(\zeta',G)d\zeta' + G^2\frac{\partial}{\partial G}\int_0^\zeta F(\zeta',G)d\zeta'\right]\Delta b\frac{dh_0}{dt} + \\ &+\left[\zeta F - (1+G)\int_0^\zeta F(\zeta',G)d\zeta' - G(1-G)\frac{\partial}{\partial G}\int_0^\zeta F(\zeta',G)d\zeta'\right]\Delta b\frac{d\Delta h}{dt}\ . \end{aligned} \tag{8.6}$$

Equations (8.4) and (8.5) are the two ordinary differential equations for three unknowns: h_0, Δh and Δb. Additional relation is needed to close the problem. It can be derived from the consideration of the turbulent kinetic energy budget within the bulk of two adjacent layers, mixed layer and capping inversion layer.

For this purpose we may employ the balance equation for turbulent kinetic energy

$$\partial e / \partial t = B - \partial\Phi / \partial z - \varepsilon\,, \tag{8.7}$$

where e is the turbulent energy per unit of fluid mass, ε is its viscous dissipation rate, and Φ is the vertical flux of energy.

We adopt the hypothesis of similarity of convective regime considered, which states that the basic turbulence parameters, being normalized by the length scale $h_0+\Delta h$ and the velocity scale $w_* = \left[B_s(h_0+\Delta h)\right]^{1/3}$, cease to depend on time in their explicit form and depend on it only through these scales, i.e., they become universal functions of the dimensionless height $z/(h_0+\Delta h)$. Hence, the vertical profiles of turbulent energy and its dissipation rate can be presented in the form

$$e = w_*^2 F_e\left(\frac{z}{h_0+\Delta h}\right), \qquad \varepsilon = \frac{w_*^3}{h_0+\Delta h}F_\varepsilon\left(\frac{z}{h_0+\Delta h}\right), \tag{8.8}$$

where F_e and F_ε are the universal functions satisfying boundary conditions $F_e(1)=F_\varepsilon(1)=0$.

The employed closure hypothesis is similar to the one proposed by Deardorff (1970a,b), that has been widely used in so-called zero-order-jump models of the convective boundary layer, of which the most comprehensive was suggested by Zilitinkevich (1991). Instead of quite arbitrary height $\bar{h}$ within the limits of the IL (usually close to the height of the buoyancy flux minimum), employed by these models, we use $h_0+\Delta h$, i.e. the depth of the whole turbulized zone, as an appropriate length scale. The height $\bar{h}$ is suitable for scaling the turbulence parameters in the major part of the convective boundary layer, but not within the IL (Zilitinkevich, 1991). In Section 8.2.3 (Figures 8.3 and 8.4) we shall see that utilisation of $h_0+\Delta h$ allows to decrease the range of empirical estimates of the universal functions for e and ε precisely near the boundary layer top.

Termwise integration of (8.7) over z from 0 to $h_0+\Delta h$ gives the following entrainment rate equation:

$$\frac{10}{3}C_e\left(1+\frac{\Delta h}{h_0}\right)^{2/3}(E_h+E_\Delta)=(1-2C_\varepsilon)-2C_\varepsilon\frac{\Delta h}{h_0}-\left(1-2C_{bb}+2G\frac{dC_{bb}}{dG}\right)\left(\frac{\Delta h}{h_0}\right)^2+$$
$$+2\left(C_b-GC_{bb}+G^2\frac{dC_{bb}}{dG}\right)\frac{\Delta h}{h_0}Ri_bE_h+$$
$$+2\left[C_b-2C_{bb}-GC_{bb}-G(1-G)\frac{dC_{bb}}{dG}\right]\frac{\Delta h}{h_0}Ri_bE_\Delta-$$
$$-\frac{4}{3}C_e\left(1+\frac{\Delta h}{h_0}\right)^{5/3}De-2\frac{\Phi(h_0+\Delta h)}{B_sh_0}, \qquad (8.9)$$

where $E_h=(B_sh_0)^{-1/3}dh_0/dt$ is the dimensionless entrainment rate; $E_\Delta=(B_sh_0)^{-1/3}d\Delta h/dt$ is the dimensionless rate of changes of Δh; $Ri_b=B_s^{-2/3}h_0^{1/3}\Delta b$ is the Richardson number based on the buoyancy increment across the IL; $De=B_s^{-4/3}h_0^{2/3}dB_s/dt$ is the non-stationarity parameter introduced in Deardorff et al. (1980) [following Zilitinkevich (1991) we call it the Deardorff number]; $C_e=\int_0^1 F_e(x)dx$ and $C_\varepsilon=\int_0^1 F_\varepsilon(x)dx$ are dimensionless constants; $C_{bb}=\int_0^1 d\zeta\int_0^\zeta F(\zeta',G)d\zeta'$ is dimensionless function of G; $\Phi(h_0+\Delta h)$ is the energy flux at the boundary layer top.

The energy drain from the boundary layer occurs due to the radiation of internal gravity waves into the stably stratified layer aloft. Thorpe (1973) expressed the maximum flux of energy for maintenance of propagating waves as $\Phi_{max}=(3\pi\sqrt{3})^{-1}N^3A^2\lambda$, where A and λ are the amplitude and the length of the waves respectively. In the framework of the simple bulk approach A and λ are usually determined from the similarity arguments. Kantha (1977) assumed A to be of the order of IL thickness, while λ is proportional to the mixed layer depth, h_0. This yields

$$2\frac{\Phi(h_0+\Delta h)}{B_sh_0}=C_NRi_N^{3/2}\left(\frac{\Delta h}{h_0}\right)^2, \qquad (8.10)$$

where $Ri_N=B_s^{-2/3}h_0^{4/3}N^2$ is the Richardson number based on the buoyancy a frequency in the non-turbulent layer, C_N is a dimensionless constant.

Zilitinkevich (1991) argued that both A and λ are of the order of Δh, which gives

$$2\frac{\Phi(h_0+\Delta h)}{B_s h_0}=C_N{}' Ri_N{}^{3/2}\left(\frac{\Delta h}{h_0}\right)^3, \tag{8.11}$$

where $C_N{}'$ is another dimensionless constant. There is still no clear experimental evidence of which expression of (8.10) and (8.11) is adequate (Fernando, 1991).

An approximation of the function $F(\zeta,G)$ fitting atmospheric and laboratory data reasonably well can be obtained from geometrical arguments. The simplest polynomial obeying the conditions (8.2) reads

$$F(\zeta,G)=\left(\frac{3}{2}G-12+30C_b\right)\zeta^2+\left(28-4G-60C_b\right)\zeta^3+\left(\frac{5}{2}G-15+30C_b\right)\zeta^4. \tag{8.12}$$

The dimensionless constants C_e, C_ε, C_N or $C_N{}'$, and the shape factors $C_b(G)$ and $C_{bb}(G)$ in the above equations are to be determined from data of measurements (atmospheric and laboratory), and large-eddy simulations of convection.

In the following section the results of estimating the values of the above parameters from the empirical data, and the comparison of the calculated characteristics of the inversion layer with laboratory measurements will be presented.

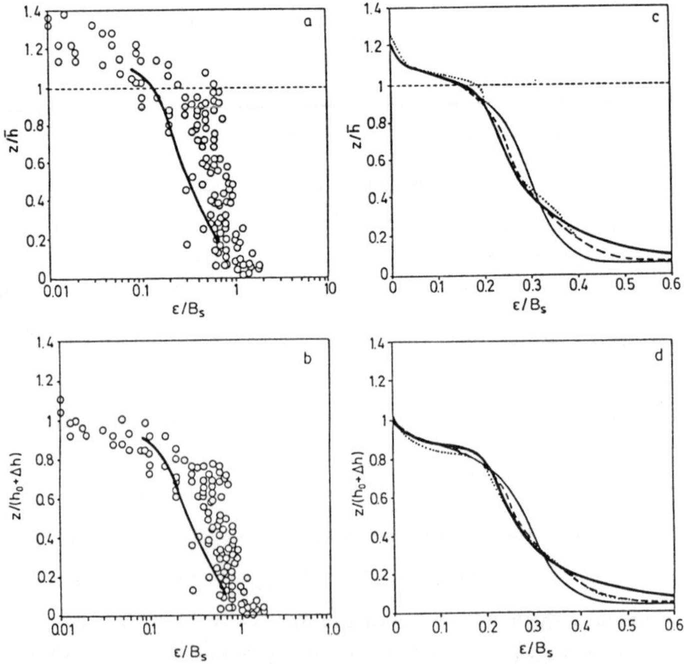

Figure 8.3. Empirical and LES data on dependence of dimensionless dissipation rate of turbulent kinetic energy, ε / B_s, on dimensionless height: (a), (c) - $z/\bar{h}$; (b), (d) - $z/(h_0+\Delta h)$. In (a) and (b) the points are the data from measurements in the atmosphere and the ocean (Caughey and Palmer, 1979; Lenschow et al., 1980; Shay and Gregg, 1986), the lines are from Deardorff and Willis (1985) laboratory experiments. In (c) and (d) the heavy solid lines show Mason's (1989) results, the solid lines are from Moeng's (1984) simulations; calculations by Nieuwstadt (1990), and Schmidt and Schumann (1989) are shown by dotted and dashed lines respectively. When the profiles of turbulent buoyancy flux were not given, the depth of the boundary layer, $h_0+\Delta h$, was determined as the crossover height of the best fit curve.

8.2.3. Model parameters and calculation results. The constants C_e and C_ε were evaluated by integrating the functions F_e and F_ε obtained from measurements and large-eddy simulations (LES), over the whole turbulized zone. Summarizing the data from Figures 8.3-8.4 gave the values C_e=0.3 and C_ε=0.3 as most appropriate for the shear-free case.

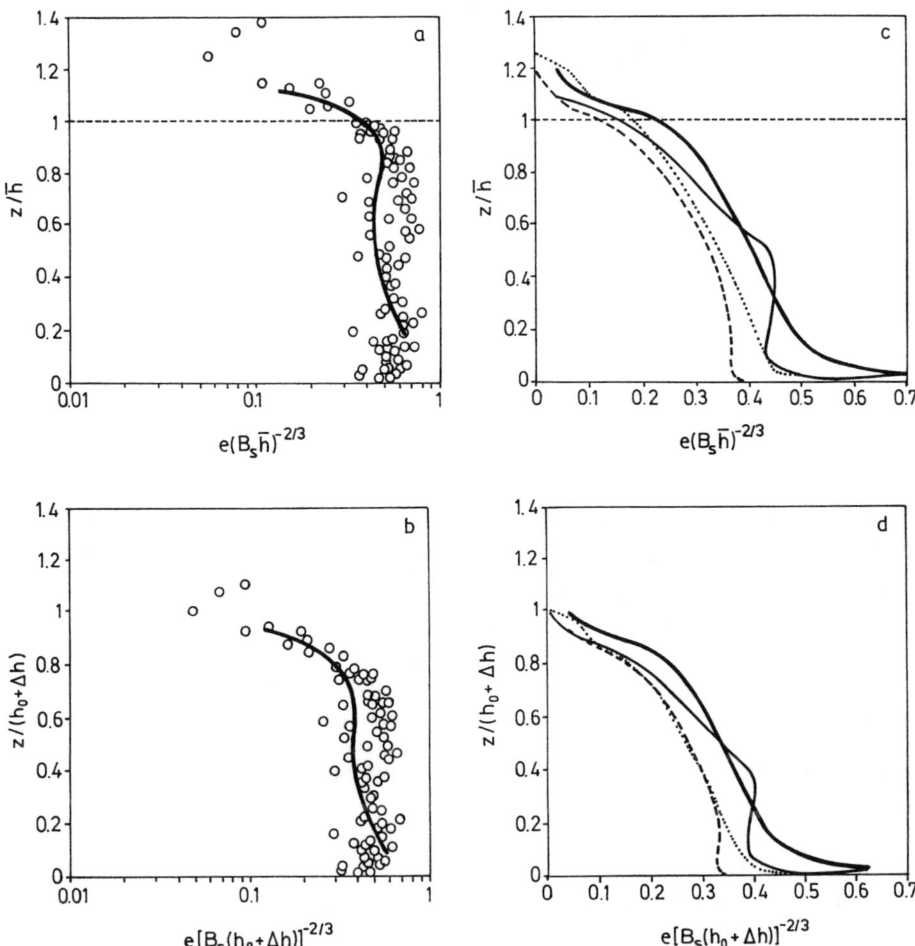

Figure 8.4. Vertical profiles of turbulent kinetic energy from measurements and LES: (a), (c) - the energy is scaled by $\left(B_s\bar{h}\right)^{2/3}$, the height is scaled by $\bar{h}$; (b), (d) - the scales are $\left[B_s(h_0+\Delta h)\right]^{2/3}$ and $h_0+\Delta h$, respectively. In (a) and (b) the points are the atmospheric data of Caughey and Palmer (1979), and Lenschow et al. (1980); the lines represent the laboratory measurements of Deardorff and Willis (1985). In (c) and (d) the LES results are shown. See Figure 8.3 for details.

We know of no direct measurements of the vertical energy flux at the IL top. Therefore, the only opportunity was to estimate dimensionless constants in (8.10) and (8.11) by comparing theoretical and empirical entrainment laws, i.e. the dependencies of the entrainment rate upon the Richardson numbers specified above. This was also a way to decide between the two expressions for wave-related vertical energy flux.

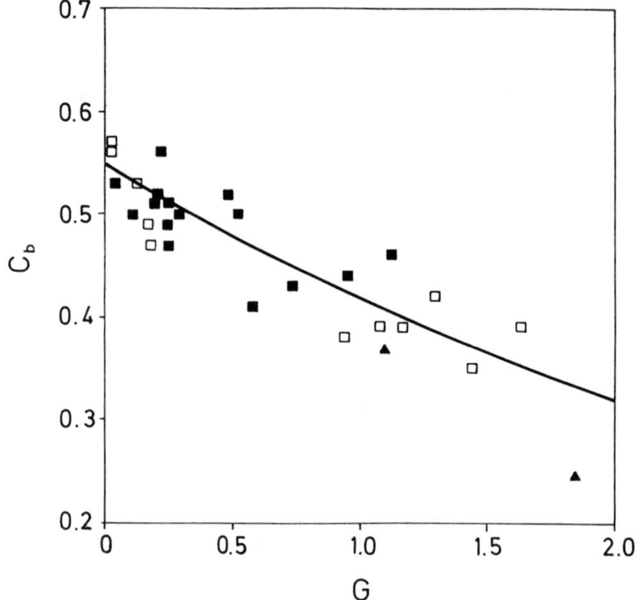

Figure 8.5. The integral shape factor C_b as function of relative stratification G. The open squares are laboratory data (Deardorff, 1979; Deardorff et al., 1980; Deardorff and Willis, 1985); the filled squares represent atmospheric data of Clarke et al. (1971), Chorley et al. (1975), Batchvarova and Gryning (1991); the triangles depict LES results of Mason (1989), and Schmidt and Schumann (1989). The line shows approximation (8.13).

Now we have to determine parameters of the function $F(\zeta, G)$. Figure 8.5 shows the dependence of C_b on G. The majority of C_b values lie between 0.35 and 0.55. The maximum value corresponding to the two-layer fluid system is 0.57. Within the observed range of G variations exponential function proposed by Deardorff (1979)

$$C_b = 0.55\exp(-0.27G) \tag{8.13}$$

fits the empirical data fairly well.

According to (8.12), function $C_{bb}(G)$ in equation (8.9) is

$$C_{bb} = \frac{1}{120}G - \frac{1}{10} + \frac{1}{2}C_b\,. \tag{8.14}$$

The empirical dependence of C_{bb} on G is shown in Figure 8.6 together with the curve calculated from equations (8.13) and (8.14). The scatter of points about the curves in Figures 8.5 and 8.6 is rather small despite the different sources of the data. This justifies the use of a simple polynomial profile (8.12), al least as a first approximation.

The model developed assumes that mean velocity shear has only a minor contribution to the total energy budget. Very few atmospheric data match this assumption. Shear-free penetrative convection was thoroughly studied in the series of laboratory experiments by Deardorff et al. (1969; 1980; Willis and Deardorff, 1974; Deardorff and Willis, 1985). The most comprehensive data set was presented in Deardorff et al. (1980, hereafter referred to as DWS). It was used to test the model.

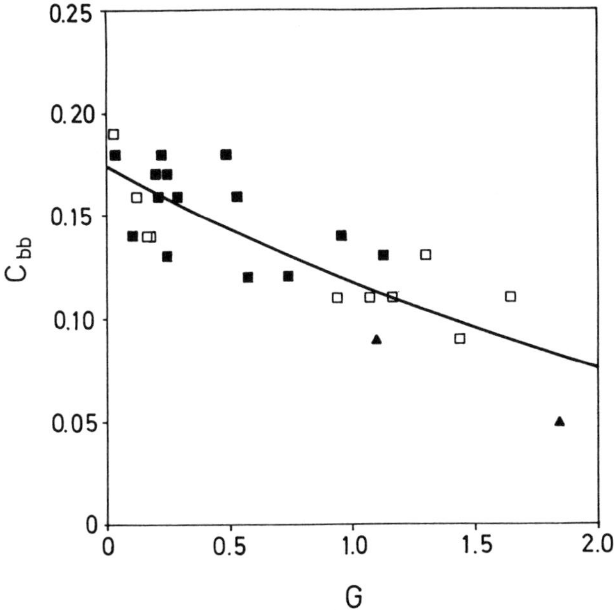

Figure 8.6. The dependence of dimensionless parameter C_{bb} on relative stratification G. Markers are the same as in Figure 8.5. The curve is drawn after equations (8.13) and (8.14).

The results of laboratory experiments and model simulations can be conveniently represented in the form of an entrainment law. A traditional way originating from the zero-order-jump approach is to express the entrainment rate and Richardson numbers in terms of $\bar{h}$, as follows:

$$\overline{E} = (B_s \bar{h})^{-1/3} d\bar{h} / dt, \quad \overline{Ri_b} = B_s^{-2/3} \bar{h}^{1/3} \Delta b, \quad \overline{Ri_N} = \frac{1}{2} B_s^{-2/3} \bar{h}^{4/3} N^2. \tag{8.15}$$

Our model allows to determine $\bar{h}$ directly from the shape of the buoyancy flux profile (8.6) as the height of the buoyancy flux minimum.

The entrainment relation for a two-layered fluid ($\overline{Ri_N}$=0) calculated by the model is depicted in Figure 8.7. It was obtained from the solution of system (8.4), (8.5), (8.9). Variations of the bottom buoyancy flux with time were neglected in the entrainment equation, as they are not important in most interesting cases including the DWS experiments (see Zilitinkevich, 1991). The initial conditions for h_0 and Δb define the initial Richardson number Ri_b. Depending upon the initial Δh value, dimensionless entrainment rate does not reveal unique dependence on Richardson number at the early stage of the buoyancy profile evolution when non-stationarity of the entrainment zone plays an important part. However, as time passes, the curves corresponding to different initial $\Delta h / h_0$ converge to a unique curve showing a quasi-equilibrium entrainment regime. The initial period of adjustment is not illustrated in Figure 8.7.

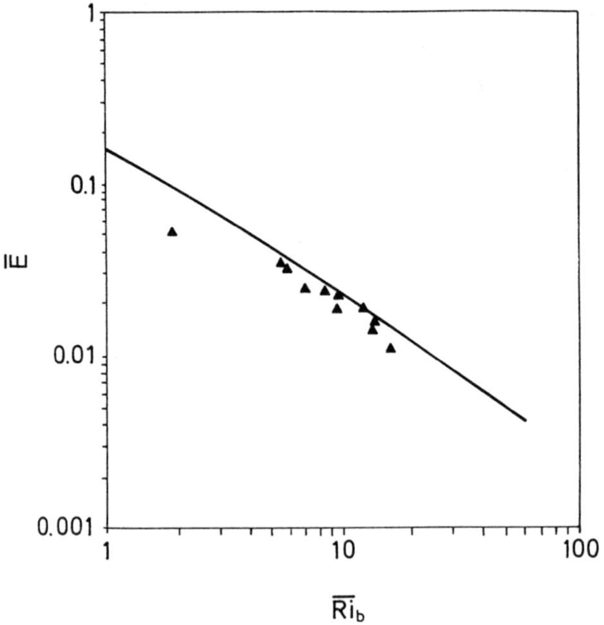

Figure 8.7. Dimensionless entrainment rate $\overline{E}$ versus Richardson number $\overline{Ri_b}$ for a two-layer fluid system. The model curve is shown by the solid line; the points are from the Deardorff et al. (1980) laboratory experiments.

In the quasi-equilibrium entrainment regime model predictions agree well with the data from two-layer fluid experiments of DWS. The dependence of $\overline{E}$ versus $\overline{Ri_b}$, shown in Figure 8.7, practically coincides with the basic relation of the zero-order model, $\overline{E} \cdot \overline{Ri_b}=const.$

Figure 8.8 displays $\Delta h/h_0$ as function of $\overline{Ri_b}$ in the two-layer fluid system. As seen from the plot, the model predicts $\Delta h/h_0$ to increase with increasing $\overline{Ri_b}$, while the DWS data show weak inverse dependence. This fact cannot be explained by the zero-order model that predicts $\Delta h/h_0$ to increase linearly at small $\overline{Ri_b}$ approaching constant at $\overline{Ri_b}$>>1. Our model results suggest that the DWS experiments inclined to study the two-layer fluid system were performed on the background of weak stable density stratification in the quiescent layer. The data from Table 1 of DWS indicate that stable lapse rate was always present at the IL top, even in experiments aimed to simulate two-layer fluid. The $\Delta h/h_0$ values from these experiments are plotted against $\overline{Ri_b}$ in Figure 8.9 with due regard to non-zero values of N^2. They conform fairly well with the theoretical curve for a linearly stratified fluid at small $\overline{Ri_b}$.

The model developed is able to reproduce the hysteresis-type relationships between the entrainment parameters, resulting from the convective boundary layer development through consecutive transition stages in a multi-layer fluid. The hysteresis entrainment zone behaviour was observed in the day-time atmospheric boundary layer (Nelson et al., 1989). This effect is qualitatively illustrated in Figure 8.10 showing the results of simulation of entrainment in a three-layer system: neutral lower layer - linearly stratified layer - neutral upper layer. The normalized IL depth grows with time during the transition period of the convective boundary layer development, whereas the Richardson number diminishes. The present model is suitable to simulate such a behaviour when the effects of wind-shear and large-scale subsidence are weak.

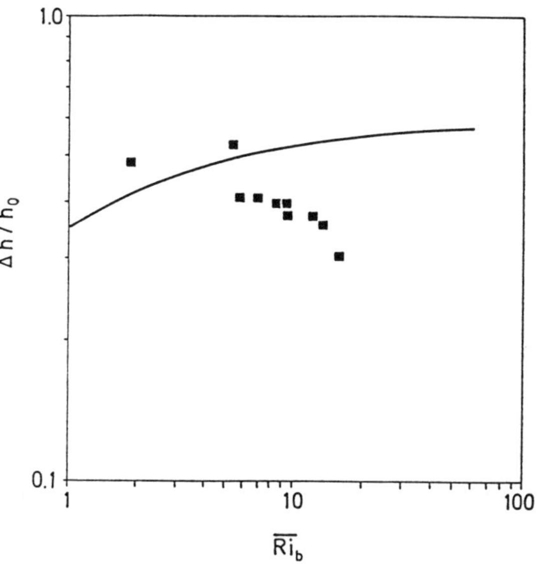

Figure 8.8. Normalized entrainment layer depth $\Delta h / h_0$ versus $\overline{Ri_b}$ in a two-layered fluid. The curve is calculated by the present model; the points are from the Deardorff et al. (1980) laboratory experiments.

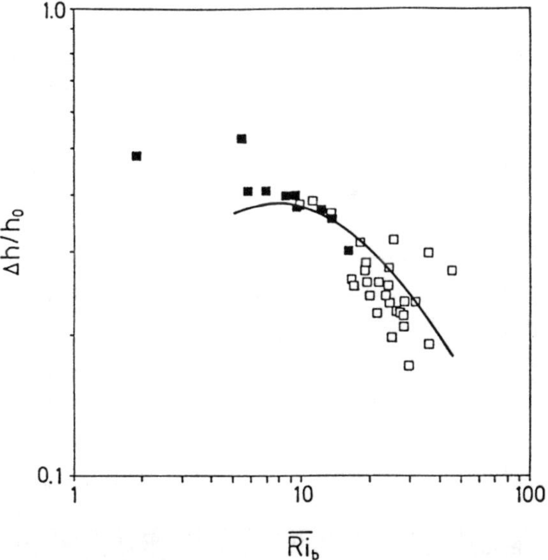

Figure 8.9. Normalized entrainment layer depth $\Delta h / h_0$ versus $\overline{Ri_b}$ in a linearly stratified fluid. The curve is calculated by the present model, employing formula (8.11) with C_N'=0.012 to parameterize the wave-related energy drain; the open squares represent the DWS data referring to a linearly stratified fluid. The DWS data from the experiments intended to treat a two-layer fluid system are also shown (filled squares).

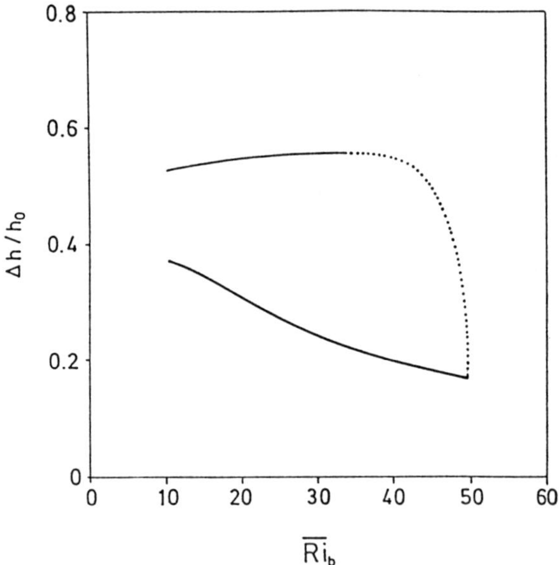

Figure 8.10. Convective boundary layer development in a three-layered fluid. The heavy solid curve corresponds to the entrainment in a linearly stratified layer. The dotted curve represents the transition period of the boundary layer development when $\Delta h / h_0$ adjusts to a regime characteristic of a two-layer system. The solid line displays quasi-equilibrium entrainment in a two-layered fluid.

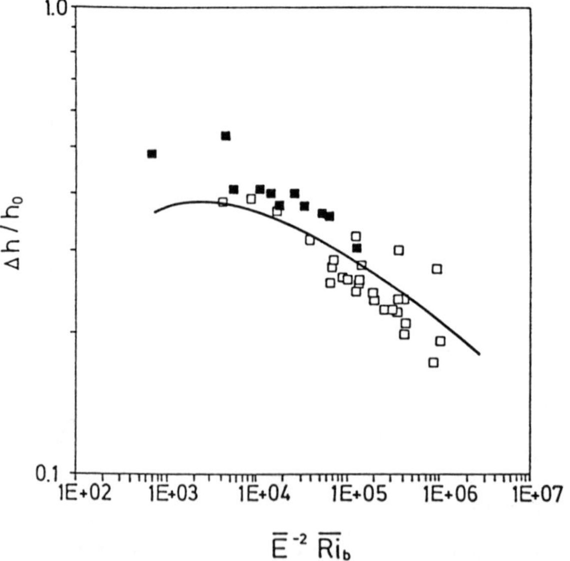

Figure 8.11. Normalized entrainment layer depth $\Delta h / h_0$ versus dimensionless combination $\overline{E}^{-2} \cdot \overline{Ri_b}$ in a linearly stratified fluid. The curve is calculated by the present model; the markers are the same as in Figure 8.9.

It is worth mentioning that the spread of data about the theoretical curve in Figure 8.9 is rather large. Basic equations of the model (8.4), (8.5) and (8.9) indicate that normalized entrainment layer depth is the function of several dimensionless parameters (dimensionless entrainment rate, Richardson numbers, the rate of changes of the IL thickness). Therefore, one can hardly expect empirical data to order well if $\Delta h/h_0$ is presented as a function of the only parameter ($\overline{Ri_b}$ in Figure 8.9). Gryning and Batchvarova (1993, personal communication) processed the data on $\Delta h/h_0$ in terms of combination $Ri_E=\overline{E}^{-2}\,\overline{Ri_b}$, which they called the entrainment Richardson number. This reduced the scatter of empirical points on the graph $\Delta h/h_0$ versus Ri_E as compared to $\Delta h/h_0$ versus $\overline{Ri_b}$. As follows from the above analysis, $/h_0$ may be strongly affected by the non-stationarity of the boundary layer. This effect is not explicitly incorporated by empirical dependence of $\Delta h/h_0$ on $\overline{Ri_b}$. The data from the DWS experiments processed in terms of $\overline{E}^{-2}\,\overline{Ri_b}$ are shown in Figure 8.11. Taking explicit account of $\overline{E}$ aligns the DWS data with the theoretical predictions of the present model.

8.3. Stable boundary layer

8.3.1. Characteristics and structure. Within the stable boundary layer the greatest static stability is near the ground. It decreases smoothly towards neutral with height (Figure 8.12). In reality not the whole SBL can be classified as the inversion layer for the inverse temperature gradients can be observed usually only in the lower portion of it. Nevertheless the entire SBL is loosely called the inversion layer or nocturnal inversion (Stull, 1988).

The wind speed in the SBL increases with height, reaching maximum close to the top of the stable layer (Figure 8.12). In this maximum the wind speed is supergeostrophic in some cases. The layer with supergeostrophic wind speed values is called the nocturnal jet. Above the jet, the wind speed and direction smoothly change to geostrophic. Beneath the jet wind direction strongly veer with height. A variety of physical mechanisms can cause the formation of the nocturnal jet. The two main of them are baroclinicity over sloping terrain, and the inertial oscillation.

The angle between the wind direction and isobars in SBL is typically rather large close to the surface and changes rapidly in the vertical as wind approaches the geostrophic value.

While turbulent motions are suppressed by stability in SBL, the buoyancy (gravity) waves can be generated and spread in the statically stable environment. Turbulence can non-linearly interact with the waves. It is not very easy to distinguish between turbulence and strongly non-linear waves in the SBL flows.

From the point of view of turbulence intensity stable boundary layers can range from being well mixed to non-turbulent. With stable stratification the buoyancy forces are damping the turbulence, making it sporadic and patchy, what may lead to the decoupling of the boundary layer from surface forcings. Since there is very little vertical mixing in SBL, temperature, humidity, and turbulence can display the fine-scale vertical structure within this layer.

Wind shears are virtually the only sources of turbulence in the SBL. Turbulence, when averaged over long times, decreases smoothly towards the top of the SBL. In some situations, turbulence is continuous over the whole depth of the layer; in other cases it might be patchy, weak and intermittent, so that only being time averaged over a couple of hours the resulting turbulence flux might appear to be acting over the whole SBL depth.

For such continuous and contiguous turbulence cases, surface forcings can be used as the key scaling parameters for the SBL structure (Stull, 1988). In this situation stably stratified turbulent flow is assumed to be in equilibrium with the surface, and similarity theory considerations are applied to construct the models. A theory of the above type was proposed by Zilitinkevich (1989a,b). It is shown in the next section how it can be used as a basis for the calculation of the dynamic and thermal structure of the stable boundary layer.

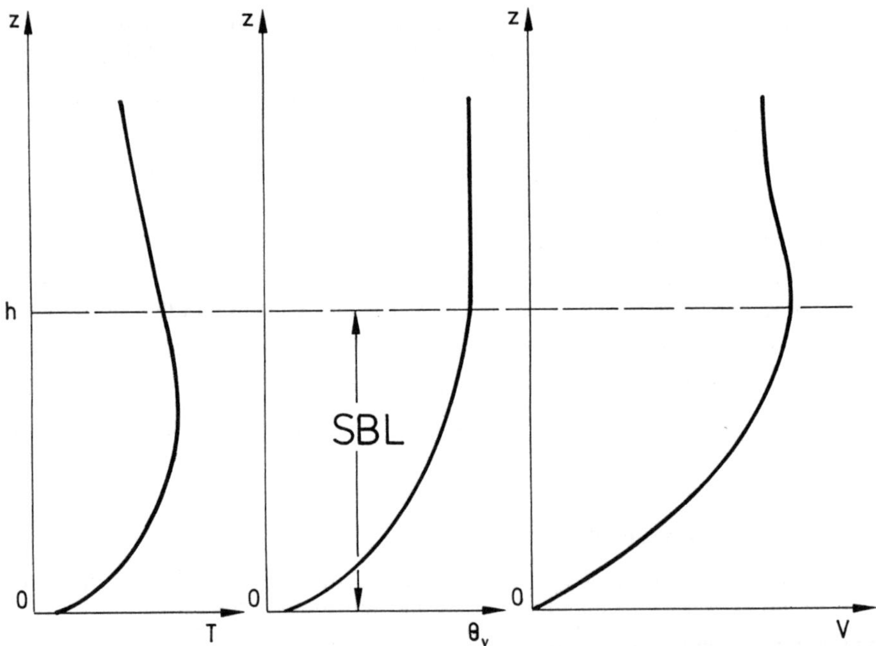

Figure 8.12. Profiles of wind speed and temperature in an idealized stable boundary layer. After Stull (1988).

8.3.2. Parameterization of wind and temperature profiles in SBL. It may be assumed that in continuously/contiguously turbulent SBL the structure of the layer at each moment of time is determined by the parameters characterizing the instantaneous situation. In case the source of buoyant forcing is located at the ground, its effect can be characterized by the near-surface value of the vertical buoyancy flux, B_s, or by the two other parameters related to it, the Monin-Obukhov length scale $L = -u_*^3 / (kB_s)$ and the non-dimensional stratification parameter $\mu = -k^2 B_s / (|f| u_*^2)$, where k is the von Karman constant (we shall use its mostly accepted value k=0.4), f is the Coriolis parameter, and u_* is the friction velocity.

At neutral stratification (B_s=0) buoyant forces do not act, at stable stratification (B_s<0) they suppress turbulence.

With the neutral stratification, the boundary layer depth is expressed by the Rossby-Montgomery formula: $h = \Lambda_0 u_* / |f|$, where Λ_0 is a non-dimensional universal constant which value can be adopted equal 0.3 (Zilitinkevich, 1989a).

With the stable stratification, the case we are interested in, the list of governing parameters determining the boundary-layer depth should be supplemented by some additional parameter related to the surface buoyancy flux.

In the case of sufficiently strong stability, when μ>>1, boundary layer depth can be expressed as (Zilitinkevich, 1972):

$$h = C_h u_*^2 |fB_s|^{-1/2}, \tag{8.16}$$

where C_h is a non-dimensional universal constant of the order of unity.

The simplest interpolation formula combining Rossby-Montgomery expression and (8.16) has the form [Zilitinkevich (1989a)]:

$$\frac{|f|h}{u_*} = \Lambda(\mu) = \left(\frac{1}{\Lambda_0} + \frac{\mu^{1/2}}{kC_h}\right)^{-1}. \tag{8.17}$$

When SBL is stationary and horizontally homogeneous, the velocity components u and v along the corresponding horizontal axes x and y satisfy the Ekman equations

$$f(v - U_g \sin\alpha) + \frac{d\tau_x}{dz} = 0, \quad -f(u - U_g \cos\alpha) + \frac{d\tau_y}{dz} = 0, \tag{8.18}$$

where z is height, $U_g \cos\alpha$ and $U_g \sin\alpha$ are components of the geostrophic wind vector (U_g is modulus of this vector, α is the angle between it and the x-axis), τ_x and τ_y are the components of the vertical turbulent flux of momentum, normalized by density. Let us direct the x-axis along the surface friction vector. The surface friction will be characterized then by the wind turn angle α and the friction velocity $u_* = \sqrt{\tau_{xs}}$. At the upper boundary of SBL, at $z=h$, the turbulence degenerates, so the vertical turbulent momentum flux reduces to zero:

$$\tau_x(h) = \tau_y(h) = 0, \tag{8.19}$$

and the wind becomes geostrophic:

$$u(h) = U_g \cos\alpha, \quad v(h) = U_g \sin\alpha. \tag{8.20}$$

Integrating equations (8.18) over z from 0 to h, taking into account the boundary conditions (8.19) and (8.20), we obtain

$$\int_0^h u\,dz = hu(h), \quad \int_0^h v\,dz = hv(h) + u_*^2 / f. \tag{8.21}$$

The SBL velocity profile at $z << h$, $z << L$ satisfies the near-surface logarithmic law:

$$u(z) = \frac{u_*}{k} \ln\frac{z}{z_0}, \qquad v(z) = 0, \tag{8.22}$$

where z_0 is the roughness parameter, and the velocity defect law at $z >> z_0$:

$$U_g \cos\alpha - u(z) = \frac{u_*}{k}\phi_u(\zeta,\mu), \quad U_g \sin\alpha - v(z) = -\frac{u_*}{k}\phi_v(\zeta,\mu)\text{sign}f, \tag{8.23}$$

where ζ is a non-dimensional height $\zeta = z/h$, and $\phi_u(\zeta,\mu)$, $\phi_v(\zeta,\mu)$ are universal functions which have to satisfy, in accordance with (8.20), the conditions $\phi_u(1,\mu) = \phi_v(1,\mu) = 0$.

Overlapping of formulas (8.22) and (8.23) in the region $z_0 << z << h$, where both hold true, gives the resistance law

$$\ln(C_g Ro) - B(\mu) = \sqrt{\left(\frac{k}{C_g}\right)^2 - A^2(\mu)}, \quad \sin\alpha = \frac{-A(\mu)}{k} C_g \text{sign}f, \tag{8.24}$$

where $Ro = U_g / (|f| z_0)$ is the surface Rossby number, $C_g = u_* / U_g$ is the geostrophic drag coefficient, $A(\mu)$ and $B(\mu)$ are non-dimensional functions of μ.

According to (8.17), (8.22-8.24) the velocity profile in the SBL can be represented as

$$u(z) = \frac{u_*}{k}\left[\ln\frac{z}{z_0} + f_u(\zeta,\mu)\right], \quad v(z) = -\frac{u_*}{k} f_v(\zeta,\mu)\text{sign}f, \tag{8.25}$$

where $f_u(\zeta,\mu)=-B(\mu)-\ln[\Lambda(\mu)\zeta]-\phi_u(\zeta,\mu)$ and $f_v(\zeta,\mu)=A(\mu)-\phi_v(\zeta,\mu)$ are functions of two dimensionless arguments, satisfying the conditions

$$f_u(0,\mu)=f_v(0,\mu)=0, \tag{8.26}$$

$$f_u(1,\mu)=-B(\mu)-\ln\Lambda(\mu), \quad f_v(1,\mu)=A(\mu). \tag{8.27}$$

With due regard to these conditions we can approximate functions f_u and f_v by the following quadratic polynomials:

$$f_u(\zeta,\mu)=b(\mu)\zeta+b^*(\mu)\zeta^2, \quad f_v(\zeta,\mu)=a(\mu)\zeta+a^*(\mu)\zeta^2, \tag{8.28}$$

whose coefficients $a(\mu)$, $a^*(\mu)$, $b(\mu)$, $b^*(\mu)$, according to (8.21), are connected in pairs

$$b^*(\mu)=-\frac{3}{2}-\frac{3}{4}b(\mu), \quad a^*(\mu)=\frac{3k}{2\Lambda(\mu)}-\frac{3}{4}a(\mu). \tag{8.29}$$

The function $\Lambda(\mu)$ is known from (8.17). To find $a(\mu)$ and $b(\mu)$ we shall use one of the results of the Monin-Obukhov similarity theory for the surface layer, namely the logarithmic + linear law for the velocity profile at $z<<h$ and $L<<h$:

$$u(z)=\frac{u_*}{k}\left(\ln\frac{z}{z_0}+\beta_u\frac{z}{L}\right), \quad v(z)=0, \tag{8.30}$$

where β_u is a non-dimensional constant, which value is about 5 (Dyer, 1974). Matching approximation (8.25) with the near-surface formulation (8.30) we are coming to

$$b(\mu)=b_0+C_h\beta_u\mu^{1/2}, \quad a(\mu)=a_0, \tag{8.31}$$

where a_0 and b_0 can be expressed through A_0 and B_0, which are the analogs of $A(\mu)$ and $B(\mu)$ for the neutral case (Zilitinkevich, 1989a):

$$b_0=6-4B_0-4\Lambda_0, \quad a_0=4A_0-6k/\Lambda_0, \tag{8.32}$$

Values of universal constants A_0, B_0 and C_h were determined by Zilitinkevich from the data of laboratory and atmospheric measurements: A_0=4.5, B_0 =1.7, and C_h=0.85.

By substituting the resulting expressions of f_u and f_v into relations (8.27) the explicit formulas for $A(\mu)$ and $B(\mu)$ can be obtained:

$$A(\mu)=A_0+\frac{3}{2C_h}\mu^{1/2}, \quad B(\mu)=B_0+\ln\left(1+\frac{\Lambda_0\mu^{1/2}}{kC_h}\right)-\frac{1}{4}C_h\beta_u\mu^{1/2}. \tag{8.33}$$

The above expressions completely specify the resistance law for the SBL provided the constants k, Λ_0, A_0, B_0, C_h, β_u, and the values of f, U_g, B_s, and z_0 are known.

From the same considerations the expression for the temperature profile in SBL is derived. According to Zilitinkevich's (1975) similarity theory, in quasi-stationary SBL the temperature defect law holds

$$\theta_h-\theta(z)=\frac{\theta_*}{k}\phi_\theta(\zeta,\mu), \tag{8.34}$$

as well as the heat transfer law

$$(\theta_h - \theta_s) / \theta_* = \frac{1}{k}\ln(u_* / f z_{0T}) - C(\mu), \tag{8.35}$$

where z_{0T} is the roughness parameter with respect to temperature, θ is the potential temperature, θ_h and θ_s are its values at the SBL upper boundary and at the underlying surface, respectively, θ_* is the temperature scale, $C(\mu)$ is the universal function of μ, and ϕ_θ is the universal function of ζ and μ.

Overlapping the above two formulas and employing the representation (8.17) for the SBL depth, we obtain the following expression for the potential temperature profile in quasi-stationary SBL:

$$\theta(z) = \theta_s + \frac{\theta_*}{k}\left[\ln\frac{z}{z_{0T}} + f_\theta(\zeta,\mu)\right], \tag{8.36}$$

where $f_\theta = -C(\mu) - \ln[\Lambda(\mu)\zeta] - \phi_\theta$ is a universal function of ζ and μ, satisfying the conditions:

$$f_\theta(0,\mu) = 0, \qquad f_\theta(1,\mu) = -C(\mu) - \ln\Lambda(\mu). \tag{8.37}$$

These conditions allow us to represent function $f_\theta(\zeta,\mu)$ in the form of a square polynomial

$$f_\theta(\zeta,\mu) = c(\mu)\zeta + c^*(\mu)\zeta^2, \tag{8.38}$$

in which coefficients $c(\mu)$ and $c^*(\mu)$ the are universal functions of μ.

Taking the first of conditions (8.37) and assuming that at the top of SBL the vertical gradient of potential temperature tends to zero, we have $c^*(\mu) = -1/2c(\mu) - 1/2$. Employing the Monin-Obukhov's logarithmic + linear law for temperature profile in the surface layer we can obtain the expression relating $c(\mu)$ to $C(\mu)$ and $\Lambda(\mu)$: $c(\mu) = 1 - 2C(\mu) - 2\ln\Lambda(\mu)$, which finally yields the representation of the function $C(\mu)$:

$$C(\mu) = C_0 + \ln\left(1 + \frac{\Lambda_0\mu^{1/2}}{kC_h}\right) - \frac{1}{2}C_h\beta_\theta\mu^{1/2}, \tag{8.39}$$

in which the recommended value of C_0 is 3.7 (Zilitinkevich, 1989a), and β_θ can be taken equal to β_u (Dyer, 1974).

In the dry-atmosphere case the expression (8.36) provides for the opportunity to replace the surface buoyancy flux in the list of external parameters by the difference of temperatures at two given layers within the SBL.

The heat transfer law combined with the resistance law gives the convenient formulation for calculating the dynamic and thermal structure of the SBL in terms of external parameters, such as geostrophic wind velocity and temperature difference across the layer.

8.3.3. Calculation procedure and examples of calculated profiles. The list of external parameters for calculating wind and temperature profiles in SBL includes:

- geographic latitude of the point in which the profiles are calculated φ;
- modulus of the geostrophic wind velocity U_g;
- roughness parameter z_0;
- air temperature at the first (lower) level T_1;
- air temperature at the second (upper) level T_2.

The Coriolis parameter is related to latitude by the well known formula: $f = 2\omega\sin\varphi$, where ω is the angular velocity of the earth. The potential temperature values at the reference levels (θ_1 and θ_2 respectively) can be obtained from T_1 and T_2 provided the elevations of the levels and reference pressure (often it is possible to prescribe p_0=1000hPa) are known.

If the potential temperature difference between the reference layers is very small, the layer should be considered as a neutral one. There is no buoyant forcing with the neutral stratification, and therefore $\mu = B_s$=0. The wind profile in this case is obtained from the system of equations (8.17), (8.25), (8.28), (8.29), (8.32), supplemented by the expressions

$$u_* = U_g k\left\{\left[\ln(h / z_0) - B_0 - \ln \Lambda_0\right]^2 + A_0{}^2\right\}^{-1/2}, \quad (8.40)$$

$$\alpha = -\arcsin\left(\frac{A_0 u_*}{kU_g}\right), \quad (8.41)$$

derived by combining (8.24) and (8.25) at $z=h$, taking into account that with neutral stratification μ=0.

With stable stratification, when $\theta_2 > \theta_1$, the temperature increment $\Delta\theta = \theta_2 - \theta_1$ can be employed for calculating the temperature scale θ_* by excluding θ_s, and z_{0T} from (8.36):

$$\theta_* = \Delta\theta k\left[\ln(z_2 / z_1) + c(\mu)(z_2 - z_1) / h + c^*(\mu)(z_2{}^2 - z_1{}^2) / h^2\right]^{-1}, \quad (8.42)$$

and evaluating the stratification parameter μ which can be expressed through θ_* and u_* as:

$$\mu = k^2 \beta \theta_* / (f u_*), \quad (8.43)$$

where $\beta = g / \overline{\theta}$ is the buoyancy parameter, and $\overline{\theta}$ is the reference value of the potential temperature.

To calculate the temperature profile using formula (8.36) we need additional equation for roughness length with regard to temperature z_{0T}.. One of the possible ways to relate z_{0T} to z_0 was suggested by Brutsaert (1982):

$$z_{0T} = 7.4 z_0 \exp\left[-2.46\left(\frac{u_* z_0}{\nu}\right)^{0.25}\right], \quad (8.44)$$

where ν is the kinematic viscosity of the air. In combination with equation (8.36) applied to each reference level for temperature, Brutsaert's formula provides the values of three parameters: θ_s, z_{0T} and θ_* [see (8.42)]. The surface temperature value is presented by the formula following from (8.36):

$$\theta_s = \theta_i - \frac{\theta_*}{k}\left[\ln\frac{z_i}{z_{0T}} + f_\theta(\zeta_i, \mu)\right], \quad (8.45)$$

where i is the number of any of the reference levels for temperature, and $\zeta_i = z_i / h$.

Together with equations (8.17), (8.25), (8.28), (8.29), (8.32), (8.36), (8.38), (8.39) and expressions for surface friction parameters u_* and α [they are derived easily from (8.24) and (8.25) in the same way as equations (8.40) and (8.41)]:

$$u_* = U_g k\left\{\left[\ln(h / z_0) - B(\mu) - \ln \Lambda(\mu)\right]^2 + A(\mu)^2\right\}^{-1/2}, \quad (8.46)$$

$$\alpha = -\arcsin\left[\frac{A(\mu) u_*}{kU_g}\right] \quad (8.47)$$

expressions (8.42)-(8.45) constitute the system for calculating the velocity and temperature profiles in the stable boundary layer.

Both systems presented above include non-linear equations. To solve them by any of iterative methods, it is necessary to prescribe the initial values of one or several parameters to be determined. In the case of neutral stratification, the value of the friction velocity can be prescribed, for example as a fraction of U_g. With stable stratification, it is convenient to define also the initial value of θ_*, taking the logarithmic law as the first approximation for the temperature profile. Starting with these two parameters we can calculate the initial values of h and μ, employing formulas (8.17) and (8.43), respectively. Substitution of μ into equations (8.17), (8.29), (8.31), (8.33), and (8.39) gives the values of μ-dependent functions and coefficients of the polynomials in the representations of the velocity and temperature profiles. This allows to calculate the new approximations for the temperature scale and the friction velocity, using equations (8.42) and (8.46). The above cycle is repeated till the convergence is achieved. After that the profiles and their supplementary parameters (α, θ_s, z_{0T}) can be calculated from the equations (8.25), (8.28), (8.36), (8.38), (8.44), (8.45), and (8.47).

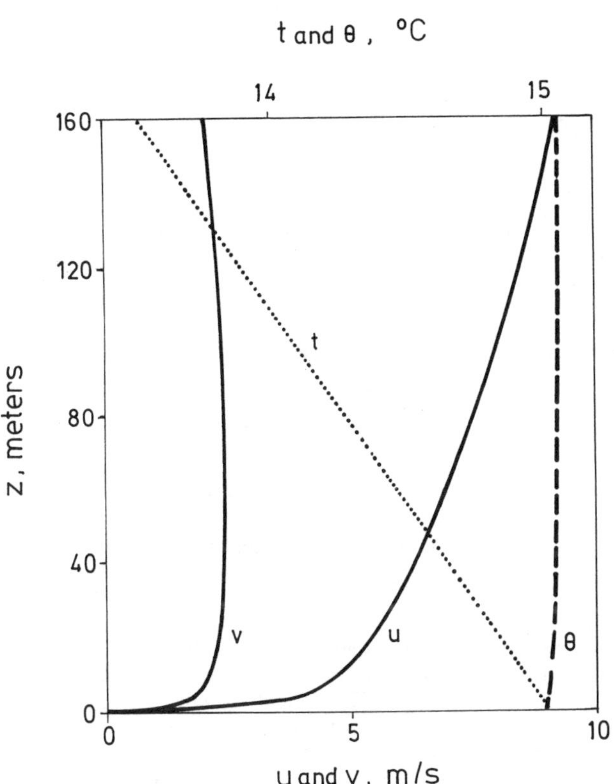

Figure 8.13. Simulated profiles of wind speed and temperature in the stable boundary layer under the conditions of very weak stratification.

The presented formulations and calculation procedure were used for simulation of the vertical structure of SBL with different rates of stability.

The examples of the calculated profiles for the boundary layer with very weak stratification, i.e., for the nearly neutral case, when μ is close to zero, are presented in Figure 8.13. The following values of the external parameters were prescribed for this case: φ=45°; U_g=10m/s; z_0=5cm; T_1=15°C (z_1=0.5m); T_2=15°C (z_2=2.0m). The resulting velocity vector was rotated by the calculated angle α to transfer velocity components into the co-ordinate system with x-axis parallel to the geostrophic wind. The temperature profile in Figure 8.13 shows typical adiabatic behaviour, the potential temperature being nearly constant with height. Boundary layer is rather deep, and wind velocity smoothly varies with height above few tens of meters within which the wind profile is close to logarithmic shape.

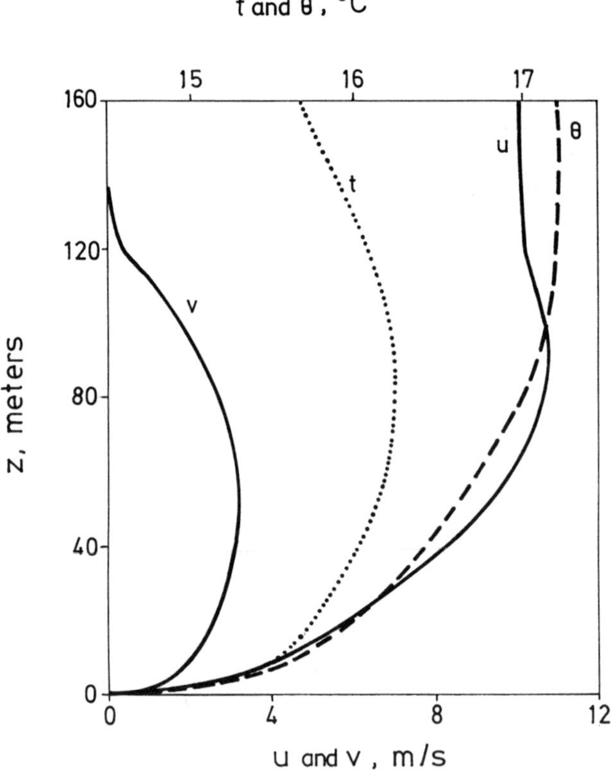

Figure 8.14. Simulated profiles of wind speed and temperature in the stable boundary layer with temperature inversion.

The velocity and temperature pattern in Figure 8.14 is quite different. Presented profiles correspond to the case when hydrostatic stability restricts the depth of the turbulent layer, and the effects of stratification on the velocity field are quite well pronounced. In the graph one can see the typical attributes of the SBL vertical structure: jet near the SBL top, wind direction strongly veering with height, low wind speed close to the surface. Still, compared to the previous plot, in the list of external parameters the only one was changed, here T_2=15.2°C (z_2=2.0m). Qualitatively calculated profiles are very similar to those presented in Figure 8.12 of Section 8.3.1 where basic features of SBL were discussed.

Acknowledgement

The author gratefully acknowledges the helpful co-operation with colleagues from the Laboratory of Fluid Mechanics of Ecole Centrale de Nantes where he started the work on this paper during his stay as an invited researcher.

References

BATCHVAROVA, E. & GRYNING, S.-E. 1991 Applied model for the growth of the daytime mixed layer. *Boundary-Layer Met.* **56**, 261-274.

BRUTSAERT, W. 1982 *Evaporation into the Atmosphere.* Reidel. 299pp.

CAUGHEY, S.J. & PALMER, S.G. 1991 Some aspects of turbulence structure through the depth of the convective boundary layer. *Quart. J. Roy. Met. Soc.* **105**, 811-827.

CHORLEY, L.G., CAUGHEY, S.J. & READINGS, C.J. 1975 The development of the atmospheric boundary layer: three case studies. *Met. Mag.* **104**, 349-360.

CLARKE, R.H., Dyer, A.J., Brook, R.R., Reid, D.G. & Troup, A.J. 1971 The Wangara experiment: Boundary layer data. Tech. Paper 19, Div. Meteor. Phys., CSIRO Australia, 363pp. [ISBN 0643 00648 6. NTIS N71-37838.]

DEARDORFF, J.W. 1970a Preliminary results from numerical integration of the unstable boundary layer. *J. Atmos. Sci.* **27**, 1209-1211.

DEARDORFF, J.W. 1970b Convective velocity and temperature scales for the unstable planetary boundary layer and for Raleigh convection. *J. Atmos. Sci.* **27**, 1211-1213.

DEARDORFF, J.W. 1979 Prediction of convective mixed-layer entrainment for realistic capping inversion structure. *J. Atmos. Sci.* **36**, 424-436.

DEARDORFF, J.W. & WILLIS, G.E. 1985 Further results from a laboratory model of the convective planetary boundary layer. *Boundary-Layer Met.* **32**, 205-236.

DEARDORFF, J.W., WILLIS, G.E. & LILLY, D.K. 1969 Laboratory investigation of non-steady penetrative convection. *J. Fluid Mech.* **35**, 7-31.

DEARDORFF, J.W., WILLIS, G.E. & STOCKTON, B.H. 1980 Laboratory studies of the entrainment zone of a convectively mixed layer. *J. Fluid Mech.* **100**, 41-64.

DYER, A.J. 1974 A review of flux-profile relations. *Boundary-Layer Met.* **1**, 363-372.

FERNANDO, H.J.S. 1991 Turbulent mixing in stratified fluids. *Annu. Rev. Fluid Mech.* **23**, 455-493.

KANTHA, L.H. 1977: Note on the role of internal waves in thermocline erosion. In *Modelling and Predictions of the Upper Layer of the Ocean* (ed. E.B.Kraus), pp.173-177. Pergamon Press.

LENSCHOW, D.H., WYNGAARD, J.C. & PENNEL, W.T. 1980 Mean-field and second-momentum budgets in a baroclinic, convective boundary layer. *J. Atmos. Sci.* **37**, 1313-1326.

MASON, P.J. 1984 Large-eddy simulation for the convective atmospheric boundary layer. *J. Atmos. Sci.* **41**, 2052-2062.

MOENG, C.-H. 1984 A large-eddy simulation for the study of planetary boundary layer turbulence. *J. Atmos. Sci.* **46**, 1492-1516.

NELSON, E., STULL, R. & ELORANTA, E. 1989 A prognostic relationship for entrainment zone thickness. *J. Appl. Meteorol.* **28**, 885-903.

NIEUWSTADT, F.T.M. 1990 Direct and large-eddy simulation of free convection. In *Proc. 9th Internat. Heat Transfer Conf., Jerusalem, August 19-24, 1990,* pp.37-47. Amer. Soc. Mech. Engrg., New York, Vol.I.

SCHMIDT, H. & SCHUMANN, U. 1989 Coherent structures of the convective boundary layer derived from large-eddy simulations. *J. Fluid. Mech.* **200**, 511-562.

SHAY, T.J. & GREGG, M.C. 1986 Convectively driven turbulent mixing in the upper ocean. *J. Phys. Oceanogr.* **16**, 1777-1798.

STULL, R.B. 1976 Mixed-layer depth model based on turbulent energetics. *J. Atmos. Sci.* **33**, 1268-1278.

STULL, R.B. 1988 *An Introduction to Boundary Layer Meteorology*. Kluwer Academic Publishers. 666pp.

WILLIS, G.E. & DEARDORFF, J.W. 1974 A laboratory model of the unstable planetary boundary layer. *J. Atmos. Sci.* **31**, 1297-1307.

THORPE, S.A. 1973 Turbulence in stably stratified fluids: a review of laboratory experiments. *Boundary-Layer Met.* **5**, 95-119.

ZILITINKEVICH, S.S. 1972 On the determination of the height of the Ekman boundary layer. *Boundary-Layer Met.* **3**, 141-145.

ZILITINKEVICH, S.S. 1975 Resistance laws and prediction equations for the depth of the planetary boundary layer. *J. Atmos. Sci.* **32**, 741-752.

ZILITINKEVICH, S.S. 1989a Velocity profiles, the resistance law and the dissipation rate of mean flow kinetic energy in a neutrally and stably stratified planetary boundary layer. *Boundary-Layer Met.* **46**, 367-387.

ZILITINKEVICH, S.S. 1989b The temperature profile and heat transfer law in a neutrally and stably stratified planetary boundary layer. *Boundary-Layer Met.* **49**, 1-5.

ZILITINKEVICH, S.S. 1991 *Turbulent Penetrative Convection*. Avebury Technical. 179pp.

Address of the author

Dr. Evgeni Fedorovich
Institut für Hydrologie und Wasserwitschaft
Universität Karlsruhe
76128 Karlsruhe, Deutschland
gg07@dkauni2.bitnet

Index